ARISTOTLE'S SCIENCE OF MATTER AND MOTION

Aristotle's Science of Matter and Motion

CHRISTOPHER BYRNE

UNIVERSITY OF TORONTO PRESS
Toronto Buffalo London

Toronto Buffalo London
utorontopress.com

ISBN 978-1-4875-0396-3

Library and Archives Canada Cataloguing in Publication

Byrne, Christopher, 1953–, author
Aristotle's science of matter and motion / Christopher Byrne.

Includes bibliographical references and index.
ISBN 978-1-4875-0396-3 (cloth)

1. Aristotle. 2. Matter. 3. Motion. I. Title.

Q151.B97 2018 117 C2018-902364-3

This book has been published with the help of a grant from the Federation for the Humanities and Social Sciences, through the Awards to Scholarly Publications Program, using funds provided by the Social Sciences and Humanities Research Council of Canada.

University of Toronto Press acknowledges the financial assistance to its publishing program of the Canada Council for the Arts and the Ontario Arts Council, an agency of the Government of Ontario.

Canada Council for the Arts | Conseil des Arts du Canada

Funded by the Government of Canada | Financé par le gouvernement du Canada

Contents

Preface

At least since Charles Darwin's letter of 1882 to William Ogle, favourably comparing Aristotle to Linnaeus and Cuvier, modern biologists have acknowledged Aristotle's contribution to their field.[1] In the case of physics, however, the situation is almost completely reversed: Aristotle is typically portrayed as the great delayer, the intellectual authority who held up the birth of physics until the seventeenth century, when the scientific revolution finally overthrew Aristotelianism.[2] Several of Aristotle's doctrines are mentioned in this context: his claim that the earth is at rest at the centre of the physical universe; his argument against the possibility of a void in nature; his view that all physical objects move naturally to, or within, a certain place in the universe. These claims certainly distinguish Aristotle's views from those of modern physics, but they were not, in the standard account, the main reason that Aristotle held up the development of physics. The primary difficulty was that Aristotle lacked any systematic account of matter and motion. He understood nature solely in terms of the specific natures that distinguish perceptible objects from one another and the specific motions and changes that these natures produce. As a result, he missed the properties common to all perceptible objects and the changes they undergo. In effect, Aristotle never considered perceptible objects simply as changeable physical objects. While he may have helped biology, he pointed physics in the wrong direction.

Against the above, this book argues that Aristotle offers a systematic account of matter, motion, and the basic causal powers that operate in all physical objects because of the matter from which they are made. In particular, it argues that the traditional view of what Aristotle says about the material causes of perceptible objects is incorrect: contrary to the traditional view, Aristotle holds that all perceptible objects are ultimately made from physical stuff of one kind or another and that this matter is responsible for many of their basic properties. Similarly, Aristotle holds that all types of change, beginning with the most fundamental kind of change – locomotion – have certain properties in common, regardless of the specific nature of the object in motion. This book also argues that Aristotle's account of the basic properties of matter and motion is non-teleological. Aristotle's use of final causes in biology and elsewhere is compatible with his non-teleological account of matter and motion, because all goal-directed change takes place in physical entities more complex than the material elements. Indeed, Aristotle's biology presupposes his physics, because goal-directed changes presuppose non-teleological changes in matter.

Given the topic of this book, its title should have been *Aristotle's Physics*. One of Aristotle's works, however, already has a title that is usually translated into English as his *Physics*. Thus, one has to distinguish between Aristotle's *Physics* and his physics. The relation between these two is similar to the relation between his metaphysics and his *Metaphysics*: just as some, but not all of Aristotle's metaphysics is found in his *Metaphysics*, so too some, but not all, of Aristotle's physics is found in his *Physics*. To make it clear that the subject of this book is Aristotle's physics and not just his book by that name, it is called *Aristotle's Science of Matter and Motion*.

Acknowledgments

I gratefully acknowledge the permission of the following publishers to reprint short excerpts from articles I have published with them:

"Prime Matter and Actuality," *Journal of the History of Philosophy* 33, no. 2 (1995): 197–224. Copyright © 1995 *Journal of the History of Philosophy*, Inc. Reprinted with permission by the Johns Hopkins University Press.

"Matter and Aristotle's Material Cause," *Canadian Journal of Philosophy* 31 (2001): 85–112, reprinted with permission by the Taylor and Francis Group, www.tandfonline.com.

"Aristotle on Physical Necessity and the Limits of Teleological Explanation," *Apeiron* 35 (2002): 19–46, reprinted with permission by Walter de Gruyter, Inc.

"Compositional & Functional Matter: Aristotle on the Material Cause of Biological Organisms," *Apeiron* 48 (2015): 387–406, reprinted with permission by Walter de Gruyter, Inc.

In addition, I wish to thank St Francis Xavier University, Antigonish, Nova Scotia, which granted me a sabbatical leave to work on this book; the external reviewers and editorial staff at the University of Toronto Press, who did much to improve this book; and the members of my family, who gave me their generous support.

ARISTOTLE'S SCIENCE OF MATTER AND MOTION

The Case against an Aristotelian Physics

The case against an Aristotelian physics is simply stated: physics is the science of matter, motion, and the basic causal powers that bring about change in material objects, and what Aristotle says on these topics is fundamentally misguided. The problem is not that he gets some parts wrong; the problem is that he lacks the basic building blocks required for such a science. A disorganized collection of outlandish claims about the behaviour of material objects does not count as physics.

The argument for this view draws on Aristotle's own requirements for a properly constituted science. Every science, he argues, attempts to establish what is responsible for what in its subject area.[1] In order to do this, it attempts to discover how the things it investigates are constituted and why they behave as they do. In making these inquiries, every science is guided by the principle that the explanatory principles it seeks must be appropriate to its subject matter. Thus, in addition to the common axioms that govern all of the sciences, every science or cluster of related sciences has its own principles that apply to only its subject matter.[2] In the case of the empirical sciences – the sciences of perceptible objects and their changes – one such basic principle is the principle of causality, which holds that a certain kind of cause is required to produce a certain kind of effect; it is not the case that just anything can cause anything.[3]

If there is to be an Aristotelian physics, it too must satisfy these requirements. Physics attempts to understand the perceptible universe by considering the basic properties of matter and how its causal capacities account for at least some of the observed behaviour of physical objects. In order to do this, matter and its causal capacities must be distinguished from other objects of inquiry. According to a long-standing tradition, Aristotle failed to make this distinction because he never came up with a clear concept of matter in its own right.[4] Although he discussed perceptible objects, he never properly identified their strictly physical properties or the matter from which they are made.[5] In short, he never got to the physical nature of perceptible objects.

His failure here, it is argued, is grounded in a systematic mistake on his part, namely his views on the material causes of perceptible objects. A material cause is something from which something else is made; the material causes of perceptible objects are what perceptible objects are made of.[6] Understanding the material causes of perceptible objects is crucial for physics, because we typically get at the nature of matter by looking at the basic raw materials, the physical stuff, from which perceptible objects are made. Investigating the nature of matter in this way does not commit us to any particular view of matter. Although these raw materials initially appear to us to be extended in three dimensions, internally divisible, and subject to locomotion, it might turn out that matter ultimately lacks these properties, or that these are not its fundamental properties. When we think about matter as physical stuff, all we are doing is isolating one part of perceptible objects – the basic material causes that persist through their various alterations and changes – and setting aside all of their other features; the nature of matter itself remains to be investigated. Thus, if Aristotle is going to have a science of matter, he must first be able to identify the matter from which perceptible objects are made and which aspects of their behaviour are due to that matter. A fundamentally flawed understanding of the material causes of perceptible objects makes it impossible to get a clear view of matter.

The problem, then, is not that Aristotle did not think about what perceptible objects are made of, but how he thought about those material causes. More precisely, the problem lies in what he says about the material causes of natural substances, those perceptible objects that come about or exist by nature, as opposed to physical human artefacts such as statues, houses, and drinking cups. Here, it is argued, he was misled by his view that the material causes of natural substances do not have an independent set of properties and causal capacities of their own that are necessary to the natural substances made from them. He thought this way because he thought that the nature of natural substances – what is basic and permanent in them – is captured solely by their formal cause; their material cause is responsible only for what is non-essential or accidental to them.[7] In effect, the material causes of natural substances are excluded from what is basic and permanent in natural substances.

As a result, there is no room for matter in Aristotle's science of nature, because he holds that the material causes of natural substances explain nothing fundamental about them. Aristotle can have no science of perceptible objects simply as physical objects, because the material causes of the most basic perceptible objects – natural substances – are for him merely accidental to these substances. Aristotle may attribute to natural substances the properties of physical objects – say, extension in three dimensions, divisibility, and mobility – but he does not attribute these physical properties to natural substances because he thinks they are made out of matter. The fundamental mode of explanation in physics – the explanation of the behaviour of natural objects by virtue of the matter from which they are made – is barred to Aristotle.

The one important role this reading of Aristotle leaves to the material causes of natural substances is to cause natural substances to deviate from their nature. All natural substances are, at some level, composites of a formal and a material cause, and their material cause sometimes prevents the full actualization of their formal cause. In biological generation, for example, a formal cause is combined with a material cause, and sometimes the efficient cause bringing about this change is not powerful enough to "master" the raw materials; the result is a defective substance, one in which the formal cause has not been completely actualized.[8] The defective actualization of the formal cause, in turn, causes the resulting natural substance to behave in irregular, abnormal, or suboptimal ways. Moreover, even when the formal cause has been properly actualized, the material cause still makes natural substances susceptible to abnormal behaviour, because it makes them vulnerable to external interference; other perceptible objects can act upon them in a way that prevents them from exercising their distinctive capacities. Either way, the material cause is responsible only for irregular or abnormal behaviour; the regular, unimpeded behaviour of natural substances is due to their formal cause.[9] Thus, for Aristotle, there can be no laws of physics, no unchanging patterns of behaviour that all perceptible objects follow because of the matter from which they are made. Natural substances cannot all act uniformly because of their material cause, for whatever belongs to their material cause does not belong to their permanent nature.

The upshot of this view is that Aristotle lacks an account of matter as the basic movable and physical stuff from which all perceptible objects are made. As a result, he never explains any aspects of the behaviour of perceptible objects simply by virtue of the fact that they are physical objects, whatever else they may be. He also has nothing to say about the nature of motion and change as such; he considers only the distinctive changes of particular kinds of thing and how those changes are grounded in their specific nature.[10] In sum, Aristotle's account of the physical universe is fundamentally flawed, because he has nothing to say about the nature of matter or the regular changes that perceptible objects undergo because of the matter from which they are made.

Against the above, one might point to Aristotle's theory of the material elements, for the material elements are both physical bodies in their own right and the material causes out of which all other perceptible objects are ultimately made. It is argued, however, that what Aristotle says about the material elements still fits the above account of his material causes.[11] According this interpretation of Aristotle, the four sublunary elements do not persist in the natural substances generated from them and thus are not responsible for the physical nature of perishable natural substances. Indeed, the elements cannot persist in the things generated from them; if they did, these changes would be instances of alteration and not generation. For the same reason, the elements do not persist through the destruction of natural substances. Thus, the generation and destruction of natural substances reveal that the four sublunary elements are not the most fundamental material causes of perceptible objects. The four sublunary elements are themselves perishable and thus have a material cause of their own, something out of which they are made. This material cause, traditionally called "prime matter," is the ultimate material cause of all perishable objects, because the sublunary elements are made out of it and all other perishable objects are made out of the sublunary elements. According to the traditional view of prime matter, this ultimate material cause is nothing like matter in the sense of physical stuff, because it is pure potentiality, with no nature or properties of its own.[12] In particular, as pure potentiality, it is not extended or divisible in its own right. Thus, the ultimate material cause of all perishable objects is far too insubstantial to constitute anything like matter. The upshot is that, far from setting aside the deficiencies in his account of the material causes of natural substances, Aristotle's theory of the material elements only highlights its weakness: once again, perceptible objects have no common physical nature because of the material causes from which they are made.

The above is perhaps the most widespread argument against an Aristotelian physics. A second argument for this same conclusion holds that Aristotle ascribes not too little, but too much to the material causes of perceptible objects. On this view, Aristotle fails to understand matter, because he ascribes to the material causes of perceptible objects capacities and motions that belong only to the objects made out of matter, not to matter in its own right. As a result, matter and the motions of material objects become too involved with other, non-physical aspects of nature. In the case of biological organisms, for example, Aristotle seems to ascribe to their material causes causal capacities that belong, strictly speaking, only to living beings.[13] Living beings are instrumentally organized entities, whose organs, when they function properly, bring about a result that is useful or beneficial to them. To the extent that Aristotle thinks about the material causes of perceptible objects, it is argued, he understands them entirely teleologically, that is, in relation to the production of outcomes that are beneficial to the objects made from them.[14] Thus, once again the problem is not just that Aristotle's account is wrong; it is fundamentally confused, because it ascribes the wrong kind of properties to matter. In his analysis of the material world, Aristotle is guilty of a whole series of category mistakes.

Finally, it should be pointed out that some commentators have attempted to defend Aristotle against these criticisms by arguing that Aristotle does not have a bad theory of matter, because he does not have a theory of matter at all.[15] Far from being guilty of attributing the wrong kind of properties to matter, his claims are quite sensible once it is understood that he is talking only about material causes, not matter. One piece of evidence in support of this view is the enormous range of things that can act as material causes, everything from the marble of statues to the premises of syllogisms. Material causes are so diverse, it is argued, that they have only an "analogical" unity: the one thing they have in common is that they all act as material causes for something or other.[16] Even the material causes of perceptible objects are united solely by their role as material causes for perceptible objects. In short, Aristotle never thinks about matter, just about material causes.

This book argues that Aristotle has a concept of matter over and above his concept of a material cause, one that he develops when he investigates what sorts of thing act as the material causes of perceptible objects. From this investigation Aristotle concludes that, whatever else they may be, all perceptible objects are physical objects, and part of their behaviour is to be explained by that fact alone. Moreover, the properties that perceptible objects have as physical objects belong to them because of their material causes; they are physical objects because they are made out of physical material causes. These physical material causes, in turn, are all made out of physical matter of one kind or another; the ultimate material cause of every perceptible object is always matter of some kind. Thus, perceptible objects have a common set of physical attributes due to the matter from which they are ultimately made.

The same holds for motion and change. All perceptible objects, according to Aristotle, are subject to motion and change, and certain features belong to all instances of motion and change, regardless of the specific nature of the thing undergoing the change. Some of these common features are grounded in the physical attributes of the matter from which all perceptible objects are ultimately made. Thus, motion and matter are inextricably connected. Part of understanding the nature of perceptible objects, then, lies in knowing the basic features of matter and motion. It also follows that, without knowing the basic features of matter and motion, it is impossible to determine what in the behaviour of perceptible objects is due to their formal cause, for if some feature of a perceptible object is properly explained by its material cause, it is wrong to explain that feature by its formal cause.

If the general thesis of this book is correct, those who hold that Aristotle has no general account of matter and motion have misunderstood the explanatory division of labour between formal and material causes: they make formal causes responsible for too much and material causes for too little. In particular, the potentiality, privation, and irregularity attributed to the material causes of perceptible objects belong to them only extrinsically, in relation to the objects made from them; in their own right, the material causes of perceptible objects must have certain non-privative properties of their own. Given that these material properties are necessary to perceptible objects, perceptible objects all have basic, permanent properties that are due to their material cause and cannot be properly captured by their formal cause.[17]

To make this case, this book begins with Aristotle's account of the physical attributes that all perceptible objects must have simply as changeable objects and how these attributes underlie his account of efficient causality. It then turns to Aristotle's account of the material causes of perceptible objects and the physical attributes and causal powers that these material causes possess in their own right. Here it argues that the causal powers of matter are necessary both unconditionally – in the sense of operating according to fixed, invariable patterns – and also conditionally in that the objects made from them cannot function without them. It also argues that these basic causal powers operate non-teleologically. Finally, it considers the traditional view of Aristotle's material cause; here it argues that the material causes of all perceptible objects, including natural substances, must have their own, independent properties and causal capacities if they are to perform the functions assigned to them by Aristotle.

In general, the strategy of this book is to show that Aristotle's account of the material causes of perceptible objects is neither the too-empty one of pure potentiality nor the too-full one that contains extraneous, non-physical characteristics, such as those belonging to living beings. Stated differently, its goal is to restore to the material causes of perceptible objects the attributes that belong to them in their own right, and to take away from them the ones that do not.

Chapter One

Motion and Change in Perceptible Objects

The first step in our inquiry is to consider the attributes that, according to Aristotle, all perceptible objects have in common, for if all perceptible objects share a set of physical attributes due to the matter from which they are made, these physical attributes will show up among their common attributes. This chapter considers the common attributes of perceptible objects that are connected to motion and change. Chapter 2 considers the attributes connected to efficient causality. Chapter 3 considers the responsibility that the material causes of perceptible objects have for these common physical attributes.

1.1 The Common Attributes of Perceptible Objects

1.1.1 Perceptible Objects and Physical Contact

The first and perhaps most important attribute that all perceptible objects have is the capacity for physical contact with other material objects. Perceptible objects need this attribute, according to Aristotle, because it is necessary for being an object of perception at all. All perception, he argues, requires contact between the object perceived and a sense organ, whether directly or through intermediate bodies.[1] Whatever else is involved, this process always requires physical contact between extended bodies, because it is only through physical contact that perceptible bodies can act on, or be affected by, other perceptible bodies.[2] Even when a perceptible object is not in direct contact with the thing on which it is acting, there must be an intermediate body that is.[3] As Aristotle argues, something cannot act on, or be affected by, just anything else; in particular, only bodies can affect other bodies.[4]

This requirement also applies to perception. Although a sense organ typically preserves its own nature while engaged in perception, it is still altered in the sense that an external object must act on it.[5] As a result, Aristotle argues, there is no perception through a void.[6] Perceptible objects, then, must be physical bodies, because only physical bodies are capable of contact with other physical bodies, and perception requires contact between the object perceived and the sense organ perceiving it, whether directly or through intermediate bodies. Hence, both perceptible objects and sense organs must be extended bodies, capable of physical contact with other material objects.[7] It also follows that whatever lacks the attributes required to be in physical contact with other bodies is imperceptible, where "imperceptible" means not what is too small or too distant to be perceived, but what is, in principle, incapable of being perceived.

These requirements apply just as much to celestial objects – the planets and stars – as to terrestrial ones, for not only does Aristotle state the requirements for being perceptible without restriction to any particular kind of thing, but he also explains our perception of the heavenly bodies in the same way as our perception of sublunary objects, namely through physical contact.[8] Thus, sublunary and celestial objects are capable of physical contact with each other. In sum, Aristotle applies his requirements for physical interaction to all perceptible objects equally, whether made from the four sublunary elements or the fifth, celestial element. Given Aristotle's requirements for perception, all and only physical bodies are perceptible.

1.1.2 Perceptible Objects and Motion

Besides the requirements for perception, the requirements for motion also dictate that perceptible objects be physical bodies. Every perceptible object, Aristotle argues, has a natural place to which, or in the case of the heavenly bodies, within which, it naturally moves.[9] The capacity for natural motion, in turn, presupposes the capacity for motion itself, and there are features that belong to perceptible objects simply as movable objects, regardless of their natural motion.[10] Aristotle argues, for example, that every changeable object is divisible; in the case of change of place, the moving object must be divisible in each of its three dimensions of extension, for every instance of locomotion is continuous, and the motion of an object as a whole can always be divided into the partial motions of its parts.[11] In order to be divisible in each of its three dimensions, a movable object must be extended in three dimensions. Thus, every movable object must be an extended, three-dimensional body.

In addition, if something is to be capable of changing its place, it must first have a place. Strictly speaking, however, all and only movable bodies have a place.[12] Apart from movable bodies, everything else has a place only if it is found in a movable body, and as we just saw, movable bodies are always extended, physical bodies. Thus, strictly speaking, only physical bodies have places. Mathematical objects can have a position in relation to one another, but because they are not physical objects, they do not have a place.[13] In addition, movable bodies are impenetrable; it is impossible for two such bodies to occupy the same place.[14] Finally, any continuously extended physical body is infinitely divisible; that is, it is not composed of unextended parts. Thus, a physical body cannot be resolved into two-dimensional planes, a mistake, Aristotle thought, that the Pythagoreans and Platonists had made.[15] The upshot is that every perceptible object must be movable, extended in three dimensions, and occupy a place.

1.1.3 Perceptible Objects and Change

Finally, all perceptible objects must be physical bodies because of the primacy of locomotion – change of place – in relation to all other kinds of change. As we just saw, Aristotle holds that the capacity for locomotion belongs only to physical bodies. The requirements for locomotion, in turn, apply to all other kinds of change, because locomotion is the first and most fundamental kind of change, even more important than generation and destruction,[16] for all perceptible objects are capable of locomotion, but, in Aristotle's account, not all are capable of undergoing generation and destruction; only the sublunary elements and things made from them are perishable, but not the celestial bodies. Moreover, locomotion is the fundamental kind of change, even for perishable objects, for generation and destruction take place by things coming together (σύγκρισις) and going apart (διάκρισις), which require locomotion.[17] In addition, the efficient cause of any generation must move to the relevant raw materials and act upon them, which again requires locomotion.[18] Thus, generation and destruction presuppose locomotion. In sum, the general requirements for locomotion hold of all perceptible objects, whether generated or eternal. All perceptible objects are capable of locomotion, and any object capable of locomotion must be an extended physical body.

One might respond that perceptible objects are always more than just extended, movable bodies; these properties never occur alone in physical objects. While this is true, the features that belong to perceptible objects simply as extended, movable bodies can all be understood apart from the properties that differentiate them as particular kinds of perceptible object, for the requirements for motion and physical contact apply to all perceptible objects, regardless of their particular, specific nature. If anything, it is the differentiating properties of these objects – their various natural motions and specific functions – that presuppose these common attributes. Thus, the common attributes of perceptible objects as extended, movable bodies are prior in definition to their specific properties and motions: the former can be defined without the latter, but not the converse.

1.2 Motion and Natural Motion

All perceptible objects, then, are movable and must satisfy the general requirements for locomotion. Because locomotion is a type of change, all perceptible objects must also satisfy the general requirements for change. With respect to the latter, Aristotle holds that every change requires an underlying substratum and a pair of opposed properties.[19] In the process of change, one of the opposed properties is lost by the substratum and the other gained by that same substratum, which persists throughout the change. In effect, change is a process of recombination in which a persisting substratum exchanges one of its properties for an opposed one. Thinking about change in this way, Aristotle argues, preserves the general features that belong to all types of change. To begin with, when things change, they do not change into just anything else: the opposed properties that represent the beginning and end points of a change belong to the same genus. Hot and cold, for example, are both temperatures; up and down are both locations. In addition, the substratum persisting through a change satisfies the requirement that nothing comes to be from nothing, a maxim, he argues, that holds for every instance and type of change.[20] Moreover, the persisting substratum makes it possible for something to change from one state to another; if there were no persisting substratum, one object would simply be replaced by another.[21] Thus, change of any kind always involves sameness as well as difference. To avoid contradiction, what stays the same must be different from what does not. As a result, every changeable object is something composite; it is, he says, one in number, but two in kind.[22]

As we saw in section 1.1, the substratum persisting through locomotion must be a physical body, extended and divisible in three dimensions, and the contraries gained or lost are locations. These requirements are entirely neutral with respect to the specific nature of the perceptible object in motion; they also hold regardless of how the locomotion was caused, whether naturally or violently. Thus, contrary to the view sometimes expressed, Aristotle thinks about motion as such, independent of natural motion.[23]

One might respond that even if Aristotle thinks about locomotion in general, apart from natural motion, he still attaches inappropriate requirements to it. Two features, in particular, are important here, having to do with the efficient and final causes of motion. With respect to the former, it is argued that Aristotle never thinks about motion independent of its efficient cause.[24] This objection is often raised in the context of the debate about whether Aristotle's account of motion is compatible with inertial motion.[25] Aristotle holds that a body in motion requires a mover conjoined to it, either directly or through an intermediate body, which applies a force to it to keep it in motion.[26] This claim has been held to entail that not only does inertial motion never take place, in Aristotle's account, but is incompatible with motion itself: motion always involves a change in the moving body and requires, therefore, a causal agent acting on it to produce that change.[27] If motion is inconceivable apart from a causal agent acting simultaneously upon the body in motion, then inertial motion is inconceivable. Similarly, with respect to final causes, it is argued that Aristotle never thinks about motion as such, but only teleological motion, that is, motion directed to a final cause.[28]

Below it is argued that both of these claims are false. Commentators have failed to distinguish between what Aristotle says about motion and place, on the one hand, and natural motion and natural place, on the other. This division follows the traditional distinction between kinematics – the science of motion – on the one hand, and dynamics – the science of the causes of motion – on the other. As we shall see below, Aristotle has a kinematics that is conceptually prior to and independent of his dynamics. As a result, his definition of motion is compatible with inertial motion. It is only with respect to the natural motions of perceptible objects that inertial motion is ruled out.

1.2.1 Motion and the Definition of Change

At the beginning of Book III of the *Physics*, Aristotle says that in order to understand nature, one must also understand motion and change.[29]

For, he argues, nature should be understood principally as a certain kind of efficient cause: the nature of naturally occurring objects lies primarily in the internal, inseparable capacities that they have to move or change themselves.[30] To understand an efficient cause, however, it is necessary to understand the effect that it produces.[31] Thus, in order to understand nature, in the sense of an internal efficient cause, we must understand the effect it produces: motion and change; the science of nature depends upon the science of motion and change, not the converse. Indeed, the science of motion and change must be conceptually prior to the science of nature, because not every motion or change is natural: some changes are violent, that is, contrary to the nature of the thing being moved or changed; some are artificially produced; some are, in and of themselves, neither according to the nature of the changing object nor contrary to it.[32] As we saw in section 1.1.2, there are many features of motion and change that are independent of the efficient causes of motion and change; thus, these features are independent of nature in the sense of an internal efficient cause of change.

The conceptual priority of change to natural change is confirmed by Aristotle's definition of change. Change of any kind, he says, is the coming to be actual of the potential, insofar as it is potential.[33] According to this definition, change is what happens when an object makes the transition from being potentially in a certain state to being actually in that state. Change cannot be understood simply in terms of what is potential, for at that point the change has not begun; it also cannot be understood simply in terms of what is actual, for at that point the change is finished.[34] Change, then, is what happens between these two states. The potential and actual states between which a change takes place are opposites belonging to the same genus of properties, and the object undergoing the change is the substratum persisting during the transition from one of these contraries to its opposite.[35] Because change always takes place in a substratum that is already something in its own right, the definition of change requires the qualification that something undergoes a change with respect to what it is potentially, not with respect to what it is actually; the substratum cannot change into what it already is.[36] In every change, then, something comes to be actual that previously existed only potentially, and this actualization takes place in a persisting substratum that is already something actual in its own right. In sum, change is the transition that a persisting substratum makes from one opposite to another.

This definition refers to what is happening in the object that is changing, regardless of the efficient cause producing the change.[37] Thus, Aristotle's definition of change is independent of natural change in the sense of internally caused change. In fact, the example he uses to illustrate his definition of change is a non-natural, artificially caused change, the coming to be of a house from its raw materials. More generally, his definition of change is independent of efficient causality of any kind, whether natural, artificial, or violent. This independence from efficient causality is confirmed by the order of the topics in Aristotle's *Physics*: in Book I, Aristotle gives an account of what motion and change are, apart from what causes them. This account applies to change in all perceptible objects, whether physical human artefacts or naturally occurring objects. His account of natural change does not come until later, in Book II of his *Physics*. Moreover, when he returns to the question of what motion and change are in Books III and V of his *Physics*, much of his discussion is independent of his views on nature; the motions of physical artefacts still have to satisfy the general requirements for motion and change. The same holds for his discussion of the related topics of infinity, time, and place in Books III and IV. Much of what he says on these topics holds for all perceptible and changeable objects, not just naturally occurring ones. Indeed, many of his examples involve artefacts, such as the place of a boat in a moving body of water. Moreover, Aristotle's account of motion also applies to violent motion, the motion of a perceptible object contrary to its natural motion. In sum, because not every change is natural, the distinctive feature of natural changes – having an internal, inseparable efficient cause – cannot be a necessary part of the definition of change.

Similarly, Aristotle's definition of change cannot be based on teleology, because not every change is directed to a final cause. Aristotle says that a change has a final cause only if it happens for the sake of some good or other, where the good in question must be something beneficial to the object for whose sake the change is taking place, even if that good is only what appears to a conscious agent to be beneficial.[38] Thus, changes that harm a perceptible object cannot have the good of that object as their final cause.[39] Even if such changes are beneficial for something else, that benefit does not remove the harm being done to the object in question or render a harmful change any less of a change. When one animal is killed and eaten by another animal, that change may be beneficial to one of them, but it is typically not to the other. Thus, a change in which a perceptible object is harmed does not have a final cause of its own. Strictly speaking, the harming of an object is beneficial only if that change produces another, beneficial change in that or another object: the final cause belongs only to the second, beneficial change, not to the first, harmful change.

Even when an object is not harmed, its motion can still fail to be beneficial insofar as that motion is violent, that is, contrary to its natural motion. In general, violent changes do not take place for the sake of a final cause, because by themselves they are not beneficial for the object undergoing the change; again, if they are beneficial, that benefit belongs to another change and another object. Finally, many changes cannot be explained teleologically because they are incidental to the object undergoing them: perceptible objects are constantly changing as a result of physical contact with one another, and many of the resulting changes neither harm nor benefit them because their nature is indifferent to these changes, such as a change of shape in a body of water.

In sum, it may be that every change is the actualization of a potentiality, but sometimes what is being actualized is not for the good of the object undergoing the change. Harmful, violent, and accidental changes occur all of the time and must still possess the basic features of all change. Thus, the distinctive features of natural and teleological change cannot be necessary to the definition of change.

1.2.2 *Motion and Place*

The distinction between change and natural change is clearest in Aristotle's discussion of locomotion, the most fundamental kind of change. Locomotion is a change of place. To have an account of locomotion that is independent of natural motion, Aristotle requires a concept of place that is independent of natural place. He gives such an account in Book IV of his *Physics*.[40] Here he argues that the place of a body is properly understood as its position in relation to other bodies.[41] More precisely, a place is the limit surrounding a body, the boundary where an enclosed body comes into contact with the other bodies immediately surrounding it. A place is like a container in that it surrounds its contents, but is not part of those contents.[42] It differs from containers in that it does not move and is no larger than the body it contains.[43] To capture the requirement that the place of a body not move, Aristotle says that the surrounding bodies in which the limiting boundary of a body is found must themselves be at rest, for a place is the limit with respect to which a body is either in motion or at rest, and the latter can be determined only in relation to surrounding bodies that are themselves at rest. Thus, a change in the surrounding bodies is not sufficient to establish that a body is in motion; the surrounding bodies could be in motion and the surrounded one at rest. A body is in motion only if it changes its location with respect to surrounding bodies that are themselves at rest.[44]

Not only is change with respect to the immediately surrounding bodies not sufficient to establish that a body is in motion, it is also not necessary; both the surrounded body and the immediately surrounding bodies could all be in motion. To illustrate this point, Aristotle gives the example of a boat moving with the current of a river.[45] Although the water immediately surrounding the boat remains the same, the boat is still in motion, for the surrounding water is changing its location in relation to the river as a whole, and the boat changes its location by virtue of being part of this larger, moving body of water. Thus, even if the immediately surrounding bodies are not changing in relation to the surrounded body, it does not follow that the latter is not moving; although a nail in the hull of a boat is not moving in relation to the boat, it does move with the motion of the entire boat.[46] If a body changes its location in relation to immediately contiguous bodies that are at rest, that body is moving with respect to its own place; if it is changing its location in relation to non-contiguous surrounding bodies because it is a part of another body, say, the nail in the hull of a boat, then it is moving with respect to the place of the larger body because it moves or rests together with that body.[47] Either way, if a body is changing its location with respect to external bodies at rest, it is in motion; if not, it too is at rest.

As a result, Aristotle's account allows for the possibility that a perceptible object is undergoing several different motions at once, depending on the number of different bodies to which it is connected. A nail in the hull of a boat has one location in relation to the rest of that boat and another by virtue of the location of the boat itself. Thus, the nail can be in motion or rest with respect to both of those locations; it can be moved, for example, to another part of the boat while the boat is sailing across a body of water. Similarly, water being carried in a container is at rest with respect to its container, even when that container itself is in motion.[48] A finite body has only one place of its own, the one immediately surrounding it and of equal size, but it can also be part of several other, larger bodies, each with its own place and each capable of being in motion or rest with respect to other bodies. Thus, a body can be simultaneously in motion and at rest with respect to these different locations, as well as participating in several different motions at once.

The above holds of all places and motions, regardless of whether the places belong to artefacts or naturally occurring objects; it also holds regardless of whether the motions are caused naturally or violently. As we saw, the location of a body is specified with respect to surrounding bodies that are at rest. In order to determine whether the surrounding bodies are in motion or at rest, their spatial relations to other bodies must be considered. This sets up the possibility of an infinite regress in attempting to decide which bodies are at rest and which in motion. As it turns out, in Aristotle's account, this regress can be stopped, for there is a set of fixed points to which all motions can be referred, namely the outermost sphere of the stars, within which all perceptible bodies are contained and the centre of which is also the centre of the earth.[49] It is important to realize, however, that for the purpose of determining which bodies are truly in motion and which at rest, any fixed point would do. In addition to providing such a fixed point, the outermost sphere of the stars performs a second function: it also provides a grid with respect to which all natural motions are determined.[50] In this way, the outermost limit of the physical universe determines not only the place, but also the natural place of a movable body. With respect to the former, however, the outermost sphere is not needed; any fixed point will do to determine the motion or rest of a body. It is only with respect to the latter, the motions of physical bodies to their natural places, that this sphere is necessary, for the natural motions of perceptible objects, in Aristotle's account, are determined in relation to this outermost sphere. In other words, the outermost limit of the physical universe is not a necessary part of Aristotle's account of the motions of bodies from one place to another, but only of their *natural* motions to their *natural* places.[51] If there were no such fixed limit to the physical universe, there would be no natural places and no natural motions to these places.[52] Everything that Aristotle says about place, however, would still stand.

Thus, Aristotle's discussion of place nicely illustrates the difference between motion and natural motion: part of what he says about place holds for locomotion of any kind; part has to do with the proper subset of motions that are the natural motions of physical bodies to their natural places. Every locomotion, whether naturally or violently caused, involves a change of location, and the latter is determined in relation to external bodies that are at rest. In addition to possessing all of these features, natural motions are also caused in a certain way, namely by causal powers that are located in, and inseparable from, the moving object; they are self-motions, that is, self-caused motions. The difference between motion and natural motion is the difference between motion and the causes of motion; the former is studied by kinematics, the science of motion, and the latter by dynamics, the science of the causes of motion.

This distinction between Aristotle's account of motion and his account of natural motion is compatible with his using the latter to sort out problems that arise from the former. Aristotle lacks a rigorous statement of the modern notion of an inertial frame of reference, but as we saw above, he is well aware of the difficulties involved in ascribing motion or rest to bodies. He is also well aware that the appearance of motion can be affected by the position of the observer and the observer's proximity to the object in motion. He argues, for example, that shooting stars and burning flames in the sky move with a speed comparable to that of objects thrown by us, all of which seem to move much faster than the stars, sun, and moon because they are closer to us, but are in fact moving slower than the celestial bodies.[53] In the end, he sorts out the problem of deciding which objects are in motion, in what direction, and how fast, by appealing to the natural motions of the bodies involved, part of which involves appealing to their natural places. Thus, there are aspects of the motions of perceptible bodies that can be understood only by looking at their natural motions and the internal causal powers that produce them; sometimes Aristotle uses his dynamics to sort out problems that cannot be solved by his kinematics alone. This appeal to the causal powers of perceptible bodies, however, does not negate what Aristotle says about motion and change that is independent of the natural powers producing them.

1.2.3 Motion and Inertial Motion

Aristotle's concept of motion, then, is quite compatible with inertial motion. The principle of inertia tells us what the motion of a body will be in the absence of external forces acting upon it. As such, it is not a statement about the nature of motion itself; the latter is still understood as a change of place. Rather, the principle of inertia denies that a body in motion must have a causal agent acting on it simultaneously to keep it in motion. In this sense, it is a statement about the motion of bodies in the absence of any such efficient causes. As it turns out, Aristotle's own prediction about the motion of a body in such conditions is the same as the one made by the principle of inertia.[54] In his discussion of the void, Aristotle claims that a body set in motion in a void would continue moving ad infinitum, in a straight line, unless impeded by another, larger body.[55] He goes on to say that no physical body in fact moves this way, but not because such motion is logically impossible or incompatible with motion as such. He argues, instead, that inertial motion is incompatible with natural motion: bodies moving in a void would not have the natural motions that we observe them all to have.[56] Thus, there can be no void and the world must be a plenum. His mistake here was to think that because all bodies are constantly subject to the action of external causal agents, no part of their motion is inertial. His rejection of inertial motion, however, is grounded in his account of the powers constantly acting on bodies, causing them to move or come to rest, not on the nature of motion itself.

This claim about the compatibility of Aristotle's kinematics with inertial motion is confirmed by his account of projectile motion. Aristotle thought that an external efficient cause was required to explain the motion of projectiles, because he thought that all such bodies have a natural tendency to move towards their natural place, and this natural motion was contrary to their projectile motion.[57] Projectiles are typically made out of materials that have a natural tendency to fall downward, towards the middle of the earth. In order to overcome this natural tendency, Aristotle believed that the non-downward, horizontal motion of such projectiles requires an efficient cause; without it, their horizontal motion would cease.[58] This additional causal power had to be external to these projectiles, because it would be working against the natural power already located within them that was causing them to fall to the ground. Like all perceptible objects, projectiles derive their natural motion to a certain place in the universe from the material elements from which they are predominantly made.[59] The material elements, in turn, have only one natural motion because of the simplicity of their nature.[60] Thus, unless a perceptible body is internally heterogeneous, with discrete parts working to produce a different motion, it has only one natural motion. If there is a natural power inside a simple, internally homogeneous body causing it to fall to the ground, there cannot be another natural power in that same simple body causing it to move in a different direction.

The principle of inertia, then, is not the rejection of Aristotle's concept of motion, but of his concept of natural motion. More precisely, it rejects his view that every motion is either natural or violent in the sense of having either a natural or a violent cause.[61] As we saw above, Aristotle holds that natural motion comes about through a certain kind of efficient cause, one that is necessarily located within the moving object and enables it to move itself.[62] Violent motion – the motion of a body contrary to its natural motion – is externally caused, through contact with another body that is itself moving either naturally or violently; if the external body is moving violently, that motion must, in turn, have been caused by another external body acting on it. A series of simultaneous, violently caused motions cannot go on forever; ultimately, a body must be reached that causes something else to move by means of its own internal causal power. Thus, every motion – even violent motion – must ultimately have a natural cause. As a result, Aristotle's claim that every locomotion is either natural or violent amounts to the claim that every motion is caused either by an internal, inseparable efficient cause, or by an external causal agent acting by virtue of its own natural capacity for self-motion. Stated differently, every motion ultimately has some natural cause or other. The problem with inertial motion is that, by Aristotle's principles, it is neither natural nor violent, and such motion, he thought, never occurred because every moving body is moving by virtue of either its own natural motion or the natural motion of something else.

Aristotle's discussion of the inertial motion of a body in a void is part of his argument against the ancient atomists' account of nature, in particular their claim that the behaviour of all perceptible objects is ultimately grounded in the motion of bodies through a void. Such an account is incorrect, Aristotle argues, because it makes impossible the natural motions that we observe in physical objects. These natural motions – more precisely, the natural motions of the elements and everything made out of them – are, he claims, based upon contact between bodies.[63] In the absence of that contact, no natural motion is possible. Thus, there is no natural motion in a void. The difference, then, between Aristotle's position and the one that posits inertial motion is not with respect to inertial motion as such; there is no disagreement between them as to what inertial motion is or what the motion of a body in a void will be. The difference, instead, has to do with what Aristotle says about natural motion, and natural motion is defined in relation to the causes of motion, not motion itself. Aristotle rejects inertial motion on the basis of his dynamics, not his kinematics.

Chapter Two

Efficient Causality in Perceptible Objects

According to Aristotle, then, perceptible objects have several properties in common, beginning with extension in three dimensions, divisibility, and mobility. He also holds that locomotion is prior to natural motion inasmuch as not every locomotion is natural, but every locomotion is a change of place with respect to surrounding bodies at rest, during which the body in motion traverses a continuum, that is, it does not jump from place to place without being in the intermediate places. All of these basic properties are independent of the specific natures of perceptible objects and the distinctive ways in which they move or come to rest. These properties are also independent of the causal powers that produce locomotion in perceptible objects, and of the final causes to which at least some of these motions are directed. Moreover, given that all other forms of change involve locomotion, the basic features of locomotion must be found in every instance of change, again independent of the specific nature of the changing object.

In addition to the basic properties that are found in all instances of locomotion and change, there are also, in Aristotle's account, several principles that govern all causal interactions among perceptible objects. These principles hold whether the changes involved are caused naturally, violently, artificially, or accidentally (in the sense of externally caused changes that are not contrary to the specific nature of the changing object). These causal principles come in two kinds: (1) general principles that apply to all types of efficient causality; and (2) more specific principles that apply to perceptible objects simply insofar as they are physical objects. Because the causal principles of the second kind are quantitative, non-teleological, indifferent to the distinction between natural and violent motion, and depend solely upon the physical characteristics of perceptible objects, they are best understood as Aristotle's laws of mechanics.

2.1 General Principles of Efficient Causality

Aristotle's account of the general principles of efficient causality can be usefully seen as an attempt to overcome the shortcomings of two groups of predecessors.[1] Against the first group, the Platonists and Pythagoreans, he argues that the forms and mathematical entities that they make explanatorily basic cannot in fact account for efficient causality.[2] To remedy this deficiency, he introduces two principles: first, the causal powers that produce change in perceptible objects belong only to physical causal agents, and not to immaterial entities; second, in order to produce the changes for which they are responsible, physical causal agents must have the appropriate internal structure and parts. Against the second group, the materialists, he argues that the material elements are not powerful enough on their own to cause the ordered patterns of locomotion and change that we observe in the world.[3] This shortcoming in the materialist position gives rise to two more principles: first, causal agents must in some way already possess what they transmit to the objects they move; second, the powers required to produce ordered patterns of change typically belong to causal agents that are themselves composite objects consisting of a formal and a material cause, not just to the material elements out of which these causal agents are made. These four general causal principles are considered below.

2.1.1 Efficient Causality Belongs to Physical Agents

Aristotle has an agent account of causality in that he thinks it is things that produce changes.[4] It is the builder, he says, that moves bricks and builds a house, not the art of building.[5] Another example he frequently uses is that it takes a human being to generate a human being.[6] More generally, it typically takes a biological organism of a certain kind to generate another biological organism of the same kind; thus, the soul, which is the formal cause of a biological organism, is not sufficient by itself to cause biological generation.[7] Similarly, the Platonic Ideas and other immaterial objects cannot, by themselves, produce change in perceptible objects.[8] To be sure, causal agents must have the right kind of causal capacities, but these capacities can set something in motion or bring it to rest only if they are found in a physical object, for, as we saw in section 1.1.1, perceptible objects can act on one another only through physical contact, and such contact is possible only between physical objects that are themselves extended bodies. In particular, in order to be able to interact causally, the agent and patient must have a common substratum, which in turn must be something physical.[9] In sum, efficient causality in perceptible objects always requires contact between the proximate cause and the thing being moved, and such contact is possible only between physical bodies.[10] Thus, only perceptible objects can act as efficient causes of change in perceptible objects. In this sense, the set of perceptible objects is causally closed.

2.1.2 Causation by Internal Parts

In addition to requiring causal agents to be physical objects, Aristotle also requires those agents to possess the right internal parts.[11] Causal powers do not belong to just any kind of causal agent; the agent must be made out of the right sort of materials, and those materials have to be put together in the right way. Thus, it is not the case that just anything can produce any kind of change; the right kind of agent is required, with the right causal powers, and those causal powers presuppose certain internal parts in the agent.[12]

With this requirement, Aristotle avoids the sort of vacuous explanations so nicely parodied at the end of Molière's play *Le malade imaginaire* where the "dormitive" power of opium is explained by means of its *vertus dormitiva.* The problem is not that an efficient cause cannot be specified in relation to the effect that it produces; if one is using a transmission theory of causality, as Aristotle does, then the cause and effect have to be connected inasmuch as whatever is transmitted to the patient must originally have been contained in some way in the agent. This explanation, however, becomes vacuous if it does not also specify the mechanism by which the agent is able to transmit the relevant properties to the patient and the patient is able to receive them. Otherwise, the explanation becomes tautological: the causal power of the agent simply offers a description of the changes it produces, without explaining how those changes are produced.

Aristotle is aware of the need to avoid such vacuous explanations. When discussing the explanatory role of the formal cause, he points out that explaining the behaviour of a perceptible object by appealing to its formal cause can lead to empty identity statements: a human being is a human being by virtue of having the formal cause of a human being.[13]

The problem is not that the formal causes of perceptible objects are causally irrelevant. The problem is that their causal role cannot be understood apart from the particular structure and functional parts of both the agent producing the change and the object on which it is acting. To build a house, not only must there be a builder with the correct skills, but also the right set of organs with which to exercise those skills. Only by looking at the functional parts and raw materials of perceptible objects can we discover what it is about them that gives them their relevant causal powers.

2.1.3 *Transmission Model of Causal Agency*

A third general requirement for efficient causes is that the agent must in some way already contain the effect that it produces in the object on which it acts.[14] As we saw in section 1.2.1, when an object changes, a property or attribute that previously existed only potentially in that object comes to exist actually in it; this third principle of efficient causality asserts that the property being realized, or something capable of triggering the realization of that property, must previously be found in the causal agent producing the change,[15] for efficient causality works by the causal agent transmitting a property or set of properties to the object affected by it. In the case of biological reproduction, for example, the formal cause of the progenitor is brought into being in the raw materials from which the offspring is composed; thus, it takes a human being to generate a human being.[16] Even in the case of so-called spontaneous generation, the causal agent has to impart to the raw materials the distinctive properties needed to generate the offspring, even if that happens only by chance, that is, by the accidental concatenation of external factors.[17] With respect to the other, non-reproductive kinds of change, the causal agent need only contain the property that triggers the realization of a particular property in the body on which it acts.[18] Even here, however, the triggering property typically belongs to the same genus as the one being realized. The agent causing a perceptible object to harden, for example, does not need to be hard itself; as we shall see in chapter 5, both hot and cold bodies can cause hardening in other bodies, and heat and cold are not found just in hard bodies. Nevertheless, in all types of change, only certain kinds of things can produce certain kinds of effects; in particular, the agent must either possess the property that is transmitted to another object or have the property whose transmission to another object is required to produce that effect in that kind of thing.[19]

2.1.4 *Causation by Commensurate Powers*

Finally, Aristotle argues that causal agents require the right set of causal powers to produce their effect.[20] Sometimes it appears that certain effects can be produced by chance, by agents that lack the right sort of powers. Even here, however, the appropriate causal powers must be present; chance events are not uncaused events. It is just that in chance events the requisite causal powers are found in causal agents that do not normally have them. Moreover, chance events happen only irregularly; whatever happens always or for the most part in the same way requires causal agents powerful enough to produce these results without relying on the extraordinary concatenation of external circumstances that produce chance results.

The last two causal principles, the transmission model of causal agency and the need for the appropriate causal powers, are aimed in particular against the pre-Socratic materialists. That is not to say that Aristotle completely disagrees with the earlier materialists. The transmission model of causal agency, for example, is consistent with their view that when physical bodies act on one another, the object being changed is made more like the object acting on it.[21] Aristotle's objection to the materialists' view is their claim that the causal powers of the elements are, by themselves, sufficient to produce the complex physical structures and ordered patterns of locomotion and change found in nature. Against this, he argues that such structures and patterns of change can typically be caused only by things that manifest as much order in their own behaviour. In this sense, the order to be realized must already exist in the agent.[22] Similarly, Aristotle agrees with the pre-Socratic materialists that a physical causal agent is required to bring about change in perceptible objects. On their own, however, the material elements lack the kinds of causal power required to produce the ordered patterns of change that regularly take place in nature; their causal capacities are not powerful enough to produce these effects regularly.[23] In effect, the materialists leave too much to chance. Not only is a real person required to build a house, but that person also typically requires the skills of a builder.[24] The same holds for processes as diverse as biological reproduction, the combination of the elements into various mixtures, and the setting in motion or bringing to rest of physical bodies. Similarly, the object on which the agent is acting must also possess the correct passive powers; both the agent and the patient must be the right kind of thing, with the right kinds of capacity.[25] When

Aristotle says that potential causes cannot produce actual effects, his position is not just that real effects require real causes; the causal agent must also actually possess the powers needed to produce the effect in question.[26] The material elements by themselves typically do not satisfy this requirement, because their range of powers is so limited.

In sum, in Aristotle's account, all efficient causes – and, by extension, all scientific explanations of the behaviour of perceptible objects – must satisfy four requirements: (1) changes in perceptible objects can be produced only by physical agents, whether internal or external to the object being changed; (2) causal agents, as well as the objects on which they are acting, must have the particular structure and parts required to produce the effect in question; (3) the effect produced, or a property that triggers that effect, must already be found in the causal agent, and the relevant property is transmitted to the affected object by the causal agent; (4) causal agents require the right sorts of causal powers to produce the changes for which they are responsible; where the effect is a complex physical structure or an ordered pattern of change that is regularly produced, the requisite causal powers belong only to composite physical objects, possessing a formal cause over and above that of the material elements from which they are made. In short, the agent producing a change has to have the right sort of power, in the right sort of internal structure and parts.

2.2 Mechanics and the Laws of Nature

In addition to these four general principles of causality, Aristotle sets out several causal principles that govern all interactions among perceptible objects simply insofar as they are physical objects. These principles are mathematical in that the relevant physical properties are measurable quantitatively and, within certain absolute limits, vary additively and continuously.[27] Moreover, as we shall see, the causal powers these principles describe operate mechanically in the sense that they are indifferent to the distinction between natural and violent motion, depend solely upon the basic physical properties of perceptible objects, and are independent of final causes.[28] Finally, these principles apply to all perceptible objects equally, that is, to all five elements and everything made out of them. Thus, contrary to what is often claimed, Aristotle does not have two completely separate accounts of the physical universe, one for the celestial realm and another for the sublunary realm.[29]

2.2.1 *No Action at a Distance*

Physical contact is required between the proximate mover and the thing it moves.[30] Physical contact between extended, movable bodies is the fundamental mode of efficient causality in perceptible objects; whatever else is going on when perceptible objects are set in motion or brought to rest, physical contact is required, because the proximate mover and the thing being moved must be contiguous. In particular, whenever something is set in motion, this change is caused by one body pushing or pulling on another; even changes such as being carried or turned in a different direction result from a combination of pushing and pulling.[31] Again, because locomotion is involved in all other forms of change, this principle applies to them as well.

This first principle is built into the next three:

2.2.2 *Proportion of Change to Physical Contact*

Greater contact produces greater change.[32] Given that efficient causality among perceptible objects proceeds by way of physical contact, where there is greater contact, the transmission of the effect will be faster. Thus, smaller bodies mix more easily than larger ones.

2.2.3 *Spatial Dissipation of Causal Influence*

When exercising its causal powers, a causal agent produces more change in things closer to it.[33] Because change is caused by physical contact, the greater the interference from other bodies, the less the change that will be produced by the causal agent in more distant objects. Given Aristotle's assumption that the observable world is a plenum – everywhere full of matter, without any empty spaces – the greater the distance between two bodies, the greater the number of bodies that the first body, the causal agent, will need to move in order to act upon the second body. As a result, the power of a causal agent to act on bodies not immediately contiguous to it is reduced, because it has to act on them by means of intermediate bodies.

2.2.4 *Combination of Causal Powers*

Several discrete causal powers can be combined to produce a single change. This principle complements the previous one: just as one causal agent can act on several other bodies, so too several causal agents can act on some one body to produce a single motion in that body. Aristotle explains both natural and violent motions in this way, in both natural substances and artefacts. Several meteorological phenomena, for example, are caused in this way. When vaporous exhalations rise from the earth, they can, when ignited, produce burning flames and (so-called) shooting stars in the sky, which move sideways when their natural motion upwards is countered by a motion from the upper elements resisting this motion and pushing them downwards. The resulting sideways motion is a combination of the exercise of these two powers.[34] Another example is thunderbolts, which are produced by the pressure exerted when air contracts, shooting something out under pressure, which in this case pushes the thunderbolt downward, even though hot bodies naturally tend to rise. Projectile motion works in the same way; there is just the one motion of the thrown body, but that motion is the result of combining three powers acting on it simultaneously, two external and one internal.[35] One external power is exercised by the agent initially throwing the projectile, which is then carried along by the medium through which the projectile is moving. That medium, typically air, carries the projectile in much the same way that a wave in water will carry forward a body floating on its surface. A second external power working against the motion of the projectile is the resistance offered by the medium through which it is passing, again typically air. Finally, the third power acting on the projectile is its internal, natural inclination to move to its natural place. The actual motion of the projectile is the result of the combination of these three causal powers acting on it.

In addition to the above, Aristotle has three more causal principles that regulate what is transmitted from the causal agent to the body affected by it:

2.2.5 *Necessity of Prior Motion*

Change cannot come to be from a changeless state.[36] When a causal agent exercises its power to produce a change, that causal agent must either be in motion itself or be moved by something else in motion. It is never the case that motion comes to be from a state of complete immobility. It follows from this principle that if things are in motion now, which they clearly are, this has always been the case in the past. Without an accurate measurement of the quantity of motion, it is probably too much to call this principle a statement of the conservation of motion; still, it does indicate that Aristotle's physics is about the ways in which motion is directed and ordered, not why there is motion at all. Moreover, Aristotle's claim about the necessity of prior motion for causation points to not just the eternity of motion; as we shall see in section 3.3.3, he grounds the eternity of motion, at least in part, in the eternity of matter, which in turn presupposes the conservation of matter.[37]

2.2.6 *Proportion between Cause and Effect*

A larger change is caused by a larger causal power.[38] This principle is an application of the more general principle that changes cannot be produced by just anything; the cause is always proportional to the effect, because larger changes require proportionally larger powers to produce them. In this respect, the quantity of the causal power always co-varies with the quantity of the effect: one can start with an arbitrarily small causal power and the quantity of change that it produces, and then all larger causal powers will be proportional in size to the incrementally greater effect that they produce.

This proportionality between cause and effect is, in turn, the basis for what some have called Aristotle's laws of motion.[39] These laws deal with the quantifiable relations between three factors: the causal powers that act on bodies, the amount of change that these powers produce, and the amount of time that the resulting changes take. Here Aristotle argues that the amount of change in the object being moved will be in direct proportion to the amount of power applied to it by the agent, and in inverse proportion to the resistance to that change, which comes from the quantity of the body being moved and the density of the physical medium through which it is moving.[40] Time is included in calculating the amount of power applied to the body being moved as well as the amount of change produced in that body: all other things being equal, the amount of power being applied will vary in inverse proportion to the time (as the amount of power goes up, the time required to produce the same change decreases), and in direct proportion to the amount of the change (a greater power will produce more change in the same amount of time).

All of this takes place within the limits of what Aristotle takes to be physically possible; if one group of people can move a ship, it does not follow that a smaller group will be able to move it a smaller distance because they might not be able to move it at all.[41] Indeed, he uses these proportions to argue for the existence of certain physical limits in the observable world: if these limits were removed, physically impossible results would follow. Perhaps most notorious is Aristotle's use of these proportions to argue against the existence of a void in space. It is not necessary to posit the existence of a void, he argues, because it is always possible to account for any difference in the speeds of any two bodies being moved by the same causal power by appealing to the difference in density of the resisting media through which those bodies are moving.[42] The only time that the difference in speed between two bodies could not be accounted for in this way would be if one of the bodies were moving at an infinite speed. More precisely, there is no ratio or proportion between the zero resistance to motion found in a void and the resistance offered by any other medium, but there is always a ratio between the speeds of two moving bodies. To prevent the situation where there is a finite ratio between the speeds of two bodies but not between the densities of the media through which they were moving, one of the bodies, the one moving through the void, would have to be moving at an infinite speed. His point here is not to suggest that a body moving through a void would in fact move at an infinite speed; on the contrary, precisely because the latter is impossible, he concludes that there is no void in space.[43] Rather, he is saying that the speed of a body moving through a void would have to be infinite in order to refute the claim that any difference in the speed of two bodies can be explained simply by varying the relative densities of the media through which they are moving.

Aristotle also applies this argument to free-falling bodies, that is, the natural downward motion of heavy bodies: any difference in their speeds can be explained by the difference in their size, density, and shape, as well as the difference in resistance offered by the media through which they are moving. There is no need to appeal to other explanatory factors.[44] Thus, in the case of both projectiles and naturally falling bodies, Aristotle argues that his account of perceptible bodies can explain all of the observed differences in their motions without positing the existence of a void; the latter would be superfluous. He is intent on defending this position because, as we saw in section 1.2, it is crucial to his account of the natural motions of the elements and everything made from them. It is fair to say, then, that Aristotle's account of natural motion ruled out certain possibilities for his mechanics, but then this limitation is hardly unique to Aristotle's account: mechanics is always restricted to what is physically possible.

2.2.7 *Like Produces Like*

This principle can be understood in two ways, one weaker and one stronger. The weaker version is simply the principle of causal regularity: The same cause operating on the same thing in the same way cannot produce opposite effects.[45] Fire, for example, dissolves watery liquids, so it cannot also have the opposite effect of solidifying them. Stated differently, opposites can cause only opposite effects and cannot produce the same effect.[46] The second, stronger version of this principle claims that efficient causality works by the agent making the patient similar to the agent.[47] In this sense, this requirement can be thought of as an application of the third general principle of efficient causality set out above, namely Aristotle's transmission model of causal agency; in transmitting something to the patient, the agent makes the patient more like the agent with respect to the property being transmitted.[48] Instances of natural generation are the best examples of this principle; here the patient literally becomes the same in kind as the agent.[49] In other instances, it is harder to see a similarity between the causal agent and the effect it produces. Some of these harder cases can be explained by bringing in the other causal dependencies at play in them; while there is nothing inherently dangerous about the flight of an arrow through the air, its passage through the body of an animal can be deadly because the biological functions of animals are dependent upon the spatial arrangement of their physical parts. In the case of the basic active powers in the sublunary material elements – heat and cold – the patient is literally assimilated to the agent in the sense of acquiring the property that was initially found in the agent: when a hot body makes contact with another body, the affected body becomes hotter. The effect of fire on other bodies is the pre-eminent example of this heating process, because it is the pre-eminent example of a hot body.[50] The same causal regularity holds for cold bodies.

2.2.8 *Universal Application and Experiments*

The final question to be considered with respect to these causal principles is the range of their application. The great variety of things to which Aristotle applies these mechanical principles suggests that they apply to all physical objects and all types of change. Nevertheless, Aristotle makes two distinctions, between natural and violent motion, and between the motions of celestial and terrestrial bodies, which he also considers to be important. Some commentators have argued that these two distinctions are so fundamental that Aristotle's mechanical principles cannot be understood as applying to the behaviour of all physical bodies equally: these distinctions divide the physical world up into different realms, with completely different patterns of behaviour.[51] Thus, Aristotle has no science of the physical world in general, and all of his principles of motion and causality are local and specific. Moreover, it has been argued that the first distinction, between natural and violent motion, leaves no room for physical experiments in Aristotle's science of nature: because all such experiments involve the external manipulation of bodies, the motions so produced would be classified by Aristotle as violent, at least initially.[52] As a result, it is argued, for Aristotle physical experiments tell us nothing about nature; the latter includes only naturally occurring physical objects and their natural motions, not violent motions or the behaviour of human artefacts and machines. Thus, experiments reveal nothing about the physical world as a whole.

The difficulty with this view, however, is that Aristotle does apply his mechanical principles to all motions in all perceptible objects. With respect to the natural motions of celestial and terrestrial bodies, he applies his mechanical principles to both, and the phenomena he explains by these principles are not trivial. For example, when explaining the transmission of light from the stars to the sublunary realm, he applies his first principle that efficient causality requires physical contact between the causal agent and the bodies moved by it. The light of the sun is transmitted through contact with the intervening ether to the outermost sublunary element, fire, which in turn transmits that light to the next contiguous element, air.[53] Similarly, Aristotle explains the heat produced in the sublunary world by the sun by appealing to physical processes that take place in all bodies, for in addition to the heat that naturally belongs to fire, heat can also be produced in perceptible objects through friction between moving bodies. Thus, when the eternally revolving heavenly spheres rub against the next contiguous element, fire, the fire at the point of contact and the air against which that fire, in turn, rubs are heated to the point where they ignite; that heat is then transmitted to the surface of the earth by the intervening bodies.[54] Moreover, heat is produced in the sublunary realm by the sun, as opposed to the other heavenly bodies, because of its relative proximity to the sublunary realm and the speed of its motion; the moon, though nearer, moves too slowly to produce much heat through friction, and the other planets and stars do not move fast enough or are too far away.[55]

In the above examples, Aristotle explains the behaviour of celestial bodies by appealing to the behaviour of terrestrial ones; such an inference clearly presupposes that celestial and terrestrial bodies behave in the same way with respect to these properties. He also uses several of his mechanical principles, most notably the requirement for physical contact between a causal agent and the bodies it moves, the proportion between cause and effect, and the dissipation of causal influence over distance. The amount of heat produced by friction, for example, increases in proportion to the difference in speed between the two bodies in contact, and when a moving body sets other, contiguous bodies in motion, the motion transmitted in this way declines as it is successively propagated from one body to the next. These causal patterns are found in all bodies, regardless of their natural motion or particular composition. They are also independent of any final causes that their effects may serve. Thus, these causal processes operate entirely mechanically, through direct contact between physical bodies of various sizes, shapes, and motions.[56]

With respect to the second distinction, between natural and violent motion, again we see that Aristotle applies his mechanical principles universally. As we saw, this distinction is central to the debate as to whether Aristotle's physics is compatible with scientific experiments, that is, the manipulation of physical bodies so as to observe their behaviour under controlled conditions. In fact, given the applicability of his mechanical principles to all physical objects, there is no incompatibility between Aristotle's physics and experimental science. These mechanical principles govern the behaviour of all physical bodies, whether that behaviour is caused naturally, violently, artificially, or accidentally. Aristotle, for example, appeals to an artificial (i.e., humanly caused) experiment in support of his claim that seawater is an admixture of water and other simple bodies, and not a fifth sublunary element, different from fresh water.[57] He reports, if seawater is distilled through evaporation and condensation, or filtered through a wax seal, the result is fresh water, not salt water. He also applies this technique to liquids such as blood and wine to show that they too are admixtures of simpler bodies. Moreover, he observes that ships float higher in salt water than in fresh water, which indicates that salt water arises from the mixture of denser bodies with water. In fact, not only are Aristotle's mechanical principles compatible with physical experiments, they are themselves testable experimentally: because they apply just as much to artificially created situations involving physical artefacts, they must hold there as well.

In this respect, what happens with machines applies just as much to nature.

In sum, Aristotle observes the motions of all perceptible objects – beginning with the elements and extending to everything made from them, both natural and artificial – and looks for unchanging patterns in their behaviour simply as physical objects, whatever else they may be. In so doing, he applies his mechanical principles to perceptible objects by virtue of the physical characteristics common to all bodies; whatever else they may be or do, they must follow these principles. Thus, Aristotle's mechanical principles are conceptually prior to the distinction between natural and violent motion, as well as to the distinction between celestial and terrestrial motion; his mechanical principles can be specified and observed independent of these distinctions. If satisfying these conditions is sufficient for a mechanics, then Aristotle has a mechanics.

Chapter Three

The Material Causes of Perceptible Objects

We have seen that Aristotle thinks all perceptible objects are extended, corporeal objects that are subject to various forms of change due to the physical agents that act upon them. The next step in our inquiry is to consider the material causes of perceptible objects. This chapter argues that the basic physical properties of perceptible objects belong to them by virtue of their material causes; perceptible objects are physical things because they are made out of physical material causes.

3.1 The Definition of a Material Cause

In order to understand the material causes of perceptible objects, we must first consider Aristotle's account of material causes in general. Here we need to distinguish four levels. The first and most general is his definition of a material cause (ὕλη), which he defines as that out of which something is made.[1] From his examples, it is clear that he is initially thinking of the raw materials of physical human artefacts: the bronze, silver, and bricks from which statues, drinking cups, and houses are made. His examples, however, also include parts of abstract entities, such as the propositions from which syllogisms are composed. Thus, it is not the case that every material cause is something physical. Indeed, in Aristotle's account, some material causes are clearly not physical things, such as the letters of the alphabet.[2] More generally, non-physical composite objects have non-physical material causes; in addition to the propositions of which syllogisms are made, there is also the "intelligible material cause" of which mathematical objects are made, which again is not something physical.[3] Something is a material cause, then, because something else can be made from it, not because it is something physical.

As a result, some commentators have argued that Aristotle's material cause is best understood functionally; it describes a job, rather than the thing that performs that job.[4] More precisely, it is one of a pair of correlative causes, or explanatory factors, the other being the formal cause (εἶδος). The formal cause is the defining principle that gives something its particular character or nature, and the material cause is what receives these defining characteristics. Thus, calling something a material cause does not indicate what kind of thing it is or ascribe any intrinsic properties to it. On the contrary, to be a material cause is to stand in a certain relation to something else; material causes are always the material cause *of* something. Thus, there is no such thing as a material cause *tout court*. Moreover, given that the material cause is defined functionally, there is already a clear difference between a material cause and matter: to be a material cause is to stand in an extrinsic relation to something else, the thing that is made from it, whereas to be matter in the sense of physical stuff is to possess certain intrinsic, non-relational properties, such as being extended, divisible, movable, and so forth.

The functional definition of the material cause, however, is compatible with the requirement that the material causes of certain sorts of thing have to have certain intrinsic, non-relational properties. Typically, different kinds of thing require different kinds of raw materials because their raw materials must already possess certain properties in their own right.[5] As Aristotle repeatedly says, a saw has to be made out of something hard if it is to cut wood; thus, saws are made out of iron and not wool.[6] As a result, while being a material cause entails no intrinsic properties, being the material cause of a certain kind of thing typically does. Once the general role of a material cause has been defined, we can ask what kind of thing acts as the material cause of a certain object or group of objects. In our case, we are interested in the material causes of perceptible objects.

Here again it is clear from Aristotle's examples that he does not think there is any one kind of thing that acts as the material cause out of which all perceptible objects are directly made. This diversity arises not only from the different material elements, namely, earth, water, air, fire, and ether, but also from the great variety of things that act as the material causes of perceptible objects, everything from bronze and marble to flesh and bone. Thus, even if being the material cause of a perceptible object imposes certain restrictions on the kind of thing that can perform this function, many different things can still satisfy these requirements.

Underlying this great variety, however, there is a systematic unity. Aristotle allows something that acts as a material cause for something else to have a material cause of its own.[7] The bronze, for example, out of which a bronze statue is made, is itself made out of earth and water; the bronze is the statue's proximate material cause, the material cause from which it is immediately made, and the water and earth in the bronze are non-proximate, more remote material causes of the statue. It is possible, then, for an object to have a nested series of material causes: first, the proximate material cause from which that object is immediately made; then, the more remote material causes from which its proximate material cause is made; and finally, the ultimate material cause, that is, the material cause from which that object and all of its intermediate material causes are ultimately made. As a result, not only is a material cause always the material cause *of* something, it is also the material cause of something at a certain level of that thing's composition. A material cause has to be specified not only in relation to the thing made from it, but also the level at which that thing is made from it.[8]

Given this layering of material causes, the material causes of perceptible objects are systematically connected, in that they all fit into at least one branch of an ordered series that is arranged according to what is made from them, and what they are made from. One of Aristotle's examples of such a series of material causes is earth, bone, and the hand of an animal.[9] The final material cause in these series is typically occupied by the parts of perceptible objects that are themselves complex objects, such as the organs of biological organisms. Earlier in the series, the material causes are fewer and simpler, at least inasmuch as they are, in Aristotle's view, internally homogeneous and indifferent to their shape; examples he gives here include clay, gold, bone, and stone.[10] When we think of physical stuff, we typically think of these more primitive, physical material causes. Finally, at the beginning of all these series are the five material elements. All other perceptible objects are ultimately made out of these elements, but they themselves, according to Aristotle, are not made out of a more primitive material cause capable of existing apart from them.[11] The material elements, then, constitute a further subcategory of material causes, namely, the material causes of perceptible objects that are not themselves made out of anything that can exist separately on its own. They are the most primitive kind of physical stuff.

The priority of the material elements, however, is only a qualified one. Aristotle argues that the four sublunary material elements – earth, water, air, and fire – are themselves subject to generation and destruction. Thus, these elements have a material cause of their own, something out of which they are made; this material cause is traditionally referred to as "prime matter," because it is the first and most fundamental material cause of perishable objects.[12] Aristotle does not think that prime matter exists apart from the elements; it always occurs with additional properties that make it to be one of the elements.[13] Nevertheless, prime matter can be distinguished from the four sublunary elements because it persists through the processes in which these elements are generated from one another. Thus, prime matter is the ultimate material cause of all things subject to generation and destruction. All other perishable objects are ultimately made out of the four sublunary elements, and they in turn are all made out of prime matter; it, however, is not made out of anything else. Prime matter, then, is both the proximate material cause of the four sublunary elements and the ultimate material cause of all perishable objects.

When thinking about the material causes of perceptible objects, then, these four levels have to be borne in mind: (1) material causes in general, which can be found in both perceptible and imperceptible objects; (2) the material causes of perceptible objects; (3) the most primitive, separately existing material causes of perceptible objects, the material elements; and (4) the ultimate material cause of all perishable objects, prime matter. In this and subsequent chapters, we shall be concerned principally with the second and third levels, the material causes of perceptible objects and the material elements. Because, however, a number of claims made about the elements and the material causes of perceptible objects are grounded in claims about prime matter, in chapter 4 we consider the fourth level as well.

3.2 Perceptible Matter and the Division of the Sciences

Concentrating on just the sorts of thing that act as the material causes of perceptible objects, it is clear that Aristotle thinks they are quite different from the sorts of thing that act as the material causes of imperceptible things, where "imperceptible" means what cannot in principle be perceived, not what is too small or too far away to be perceived. He says, for example, that the things studied by natural science have a perceptible material cause (ὕλη αἰσθητή), whereas the objects studied by mathematics do not.[14] Not only are the things studied by natural science perceptible, they also have a material cause that is natural, corporeal, and movable.[15] Because the objects studied by mathematics do not have this kind of perceptible and changeable material cause, mathematics is more precise than natural science.[16] The contrast Aristotle is making here is not between things that have material causes and things that do not. Rather, he is arguing that there is something distinctive about the material causes of perceptible objects, something that makes perceptible objects different in kind from the things that lack such material causes, beginning with the things studied by mathematics.

Mathematics and natural science do overlap in their subject matter inasmuch as both talk about the quantitative and geometrical properties of perceptible objects.[17] Both consider, for example, the shapes and sizes of the planets and stars. Mathematics, however, is interested in these properties only for their own sake, regardless of whether they are found in physical bodies. Indeed, not only can these mathematical properties be conceptually distinguished from the physical bodies in which they can be found, they must be so distinguished in order to be understood properly. Natural science, by contrast, is not interested in quantitative and geometrical properties in their own right, but only insofar as they help to explain the behaviour of perceptible objects. Thus, when Aristotle says that natural substances have a perceptible material cause, whereas the unchanging objects studied by mathematics do not, he does not mean by "a perceptible material cause" just whatever happens to act as the material cause of a perceptible object.[18] Rather, he already has in mind the particular kind of thing that acts as the material cause of perceptible objects and how that kind of thing makes perceptible objects different from the objects studied by mathematics.

More precisely, Aristotle claims that all of the objects examined by natural science are either one of the simple, elemental bodies, or are made of these bodies.[19] It is the presence of this kind of material cause that makes perceptible objects subject to change, and its absence that makes the objects of mathematics unchangeable and imperceptible.[20] It begs the question to say that mathematical objects are unchangeable because they lack a substratum of change; what needs to be explained is why the material causes of perceptible objects make the objects made from them subject to change, whereas the material causes of mathematical objects do not. Thus, it is not just that perceptible, changeable objects are fundamentally different from imperceptible, unchangeable ones; it is also the case that Aristotle grounds this difference at least in part in what they are made of. Nor is this difference trivial; Aristotle frequently claims that natural science must look at the material causes of natural substances as well as their formal causes.[21] He clearly thinks that what something is made of makes a difference with respect to what it is.

3.3 The Physical Requirements for Motion and Change

One might concede that Aristotle thinks of the material causes of perceptible objects as different in kind from the material causes of imperceptible and unchangeable objects, and yet still argue that Aristotle never thinks of the material causes of perceptible objects as something physical in their own right, for it is one thing to claim that physical and non-physical objects have different material causes; it is another to say that the material causes of physical objects are themselves something physical. It could be that Aristotle thinks all perceptible objects are physical objects, without thinking that they must all be made out of a physical material cause. In particular, one might argue that the physical attributes common to perceptible objects belong to them by virtue of their formal causes and not their material causes. In this case, the material causes of perceptible objects would not be responsible for the character of perceptible objects as physical objects.

In fact, as we shall see below, just the opposite is the case. Not only do all perceptible objects share certain physical characteristics, but according to Aristotle these physical characteristics belong to perceptible objects, because they belong, in the first place, to their material causes, for Aristotle's requirements for motion and causal interaction are such that not only must every perceptible object be something physical, it must also be made out of something physical. In short, every perceptible object requires a physical material cause.

3.3.1 The Substratum of Physical Interaction

As we saw in chapter 1, Aristotle holds that all perceptible objects are capable of physical contact with extended, movable bodies, and only things that are themselves physical objects are capable of this kind of contact. In addition, he says that physical interaction takes place between perceptible objects only when they differ from one another by certain contrary properties, such as one being hotter than another.[22] These contrary properties, however, cannot act on each other alone; in addition, they require a common substratum, one that is the same in kind in both of the physically interacting objects.[23] Thus, the physical interaction of perceptible objects requires not only that the affected objects be physical bodies, but they must also have a common underlying nature. This common nature, in turn, he assigns to their material causes.[24] The physical interaction between perceptible objects, then, requires not only differences in their perceptible properties, but also a common nature found in their material causes.

To emphasize the role of the material cause in the physical interaction of perceptible objects, Aristotle says that when the formal cause of a perceptible object is separated from its material cause, the formal cause is no longer capable of being affected by physical objects.[25] By way of example, he says that fire has heat in a material cause, but if heat existed apart from its material cause, it could not be physically affected. As he states it more generally, agents that do not have their formal cause in a material cause cannot be affected by other, movable objects; those that do can be affected in this way.[26] Even in those cases where the efficient cause is located in a physical object, if there is no common material cause, the agent remains unaffected by the action it performs upon something else. The medical skill found in a doctor, for example, remains unaffected by the cure the doctor performs upon a patient; the medicine consumed by the patient, however, is altered in the course of the cure it produces.[27] Thus, in order to be able to interact physically with one another, perceptible objects require a common substratum, and the material causes of these interacting objects act as the required substratum.

One might object to this conclusion by arguing that it is perceptible objects – as opposed to their material causes – that act as the common substratum required for physical contact.[28] When a child throws a ball, for example, both the child and the ball are physical objects and so satisfy the physical requirements for contact between movable bodies. This objection, however, fails for several reasons. To begin with, every perceptible object is capable of physical contact with every other perceptible object; a ball can be moved not only by a child, but also by a club, a dog, a gust of wind, or a stream of water. As we saw in chapter 1, every perceptible object is capable of being affected by every other perceptible object, and any externally caused change of properties takes place by physical contact between extended, movable bodies.[29] Thus, the common substratum that makes contact between all perceptible objects possible cannot itself be constituted by any of the contrary properties that differentiate the various kinds of perceptible objects, including the differentiating properties of the five different material elements. Rather, the substratum underlying the physical interaction of all of these different objects must be constituted by attributes that all perceptible objects share; the substratum common to all perceptible objects can be defined only by properties common to all perceptible objects.

The only intrinsic attributes that all perceptible objects have in common are the ones that belong to them as physical beings, beginning with extension, mobility, and divisibility. Thus, the physical attributes common to all perceptible objects must already belong to their common substratum; the properties that differentiate perceptible objects can work on one another only if they are found in this common substratum. The upshot is that the common substratum of all perceptible objects must be matter, that is, something physical, extended, and movable in its own right; it alone is capable of acting as the substratum underlying the physical interactions between all of the different kinds of perceptible object.

The counterclaim that perceptible objects, and not their material causes, satisfy the requirements for physical contact between movable bodies is insufficient for a second reason. As we saw above, Aristotle says that only those things that have their formal cause in a material cause are capable of physical contact with other things, whereas formal causes separated from material causes are not. Here Aristotle is making the material causes of perceptible objects responsible for their capacity to be in physical contact with other perceptible objects. These material causes, in turn, must be something physical, because only something physical is capable of this kind of contact. In other words, the formal causes of perceptible objects presuppose a physical material cause, for the formal causes of perceptible objects typically define them in ways that require the ability to interact physically with other perceptible objects, and only objects with a physical material cause are capable of doing this. Moreover, this requirement for a physical material cause applies at all levels of composition in perceptible objects. For example, not only must the proximate material cause of a statue be something physical, say, marble or bronze, but the marble and bronze, in turn, must also be made out of something physical, say, earth and water. In sum, Aristotle's requirements for physical interaction dictate that the material causes of all perceptible objects must be something physical in their own right, beginning with their proximate material cause and continuing down to their most basic constituent elements. Perceptible objects never stop being made out of matter.

3.3.2 *The Substratum of Locomotion*

A second set of arguments for the view that perceptible objects must be made out of something physical arises from what Aristotle says about locomotion. As we saw in section 1.1.2, only those things that are extended and capable of occupying a place, namely finite physical bodies, are capable of locomotion. In addition to being a physical object, however, it is also the case that for Aristotle something is capable of locomotion only if its material cause is something physical.

Evidence for this view is found the *De caelo*, where Aristotle claims that everything perceptible is found in a material cause (ὕλη).[30] This claim is a premise in an enthymeme arguing for the uniqueness of the finite world we observe, against the view that there are many, perhaps even an infinite number of other worlds. The implicit premise in Aristotle's argument is that everything found in a material cause is a spatially discrete individual, that is, something that has a unique location and cannot be found in several different places at once. It follows, then, that when Aristotle says everything perceptible is found in a material cause, he cannot mean just any material cause, for not all material causes are capable of making something to be a spatially discrete individual, such as the letters from which words are made or the premises of a syllogism. Nor can the material cause here be the "intelligible matter" that individuates particular mathematical objects of the same kind, because this kind of material cause is not enough to individuate perceptible objects, and, strictly speaking, particular mathematical objects do not have locations, but only relative positions.[31] Rather, when Aristotle says that the material cause individuates perceptible objects, the material cause here must itself be something physical, for, if something has a place, it must be a movable body, and a movable body requires a physical material cause.[32] Thus, if something has a unique location, it must, ultimately, be made out of matter of one kind or another. If all perceptible objects are particular individuals, with a unique location at any given moment, and all such individuals must, ultimately, be made out of matter, then everything perceptible must, ultimately, be made out of matter. Physical matter alone individuates, in the sense of giving a unique location to a movable, spatially discrete object.

The above is confirmed by what Aristotle says about the material cause in his discussion of place. Here he argues that the material cause of a movable object is surrounded by its formal cause in the way that points, lines, and planes spatially limit an extended body.[33] It is only something physical that can be surrounded in this way and, at the same time, underlie the properties of perceptible objects, such as being hot or cold. Although a shape defines a body in the sense of making it to be a sphere, a cube, and so forth, Aristotle still insists on the distinction between the extended matter of a movable body and its surrounding shape. It may be that every material body has certain spatial limits, but these limits do not follow from its nature as an extended body, and whatever spatial limits it does have are accidental to it as an extended body.[34] Thus, matter may be said to be infinite, in the sense of lacking any intrinsic limit.

Moreover, the property of being infinite in quantity belongs most properly to a perceptible extended continuum.[35] By virtue of being perceptible, such a continuum must also be something physical. Thus, by Aristotle's principles, it is not only the case that everything movable must occupy a unique place at every moment in time. It is also the case that a movable object will occupy a discrete location only if the object's shape encloses something extended and movable. The latter, in turn, must be something physical and is contributed to the movable object by its material cause. Thus, there is no motion without a movable body and no movable body without a physical material cause. If, in turn, every physical material cause is ultimately made out of matter of one kind or another, then there is no motion without matter.

3.3.3 *The Substratum of Generation and Destruction*

A third set of considerations, this time arising from Aristotle's account of generation and destruction, again lead to the conclusion that the material cause of a perceptible object must itself be something physical. As we saw in section 1.2, Aristotle argues that every change requires a persisting substratum and a pair of opposed properties, one of which is lost and the other gained in that change.[36] Thus, it is the substratum that is responsible for the continuity required when something loses one property and gains an opposed one. If it is to persist through this process, the substratum cannot be dependent for its own nature on the opposites gained or lost in that change. If the opposite property being lost were essential to the substratum, the substratum itself would be destroyed during this process; if the opposite property being gained were essential to the substratum, the substratum itself would be generated. In neither case would the substratum persist.[37] As a result, any property gained or lost during a change must be accidental to the persisting substratum of that change.[38] Whatever persists through a change cannot be constituted by the properties gained or lost in that change. Stated differently, the substratum of every change must be neutral with respect to the opposites gained or lost in that change; if a substratum can persist through a change in which it gains or loses a certain determination, then it must be indifferent to both that determination and its privation.

These requirements apply to essential change just as much as to accidental change, for Aristotle insists that there is a persisting substratum in generation and destruction as well.[39] It follows that the difference between essential and accidental change cannot be that in essential change what is gained or lost is essential to the persisting substratum, but accidental to the substratum in accidental change. Even in generation and destruction, what is gained or lost must be accidental to what persists. Thus, the formal cause gained or lost in an essential change must be accidental to the substratum that persists through that change.[40] The converse, however, does not hold. The persisting substratum is not accidental to the formal cause gained in generation; on the contrary, the formal cause requires a certain kind of substratum. The relation between the formal cause and the persisting substratum, then, is asymmetrical: the persisting substratum is essential to the formal cause in the sense of being necessary for its actualization, but the formal cause gained or lost in an essential change cannot be essential to the substratum persisting through that change.

In the case of generation and destruction, the persisting substratum is the material cause out of which the perceptible object is made.[41] Given that every change requires a persisting substratum whose own nature is independent of what is gained or lost in that change, the generation or destruction of a perceptible object cannot be the generation or destruction of its material cause. Rather, Aristotle argues, what is generated or destroyed is the composite object, consisting of a material and a formal cause.[42] The generation of a perceptible object, then, is not simultaneously the destruction of its material cause. Otherwise, the formal cause being actualized would have to take over the role of the persisting substratum and act both as what comes into being and as what already exists and receives the formal cause.

Instead, natural generation is similar to the production of physical human artefacts in that both take place in whatever acts as the material cause of the new object.[43] In effect, natural generation is the transmission of a formal cause from an object that already possesses it to raw materials that lack it.[44] In this process, the formal cause completes the material cause. It can do this because the formal and material causes are not opposites.[45] The opposite of a formal cause is the privation of that formal cause, not the material cause. Thus, the material cause and the privation of the formal cause are distinct.[46] This distinction allows the pre-existing raw materials to be altered during generation and yet not be destroyed. They can do this because both the privation lost in generation, and the formal cause gained, are accidental to these raw materials. Thus, the persisting material cause loses none of its own essential attributes in the generation of a natural substance; it changes only with respect to its accidental properties. The persistence of a material cause does not make a change to be an accidental one. A change is an instance of generation or destruction because a formal cause is gained or lost, not because the material cause is generated or destroyed.

In sum, if something is to act as the substratum persisting through a process in which opposed properties are gained or lost, it must be capable of having these properties in the first place. Since only certain kinds of things are capable of having certain sorts of properties, only certain kinds of things are capable of acting as the substratum persisting through the processes in which these properties are gained or lost. For every change, the right kind of substratum is required.[47] Because it is accidental to the material cause, the formal cause does not explain the nature of the material cause; on the other hand, because the actualization of the formal cause presupposes a certain kind of material cause, considering what kind of formal cause is to be actualized can tell us what characteristics its material cause must have. As we saw in section 3.3.1, everything subject to change, including generation and destruction, must be a physical object. This requirement holds for the entire course of any change, that is, a particular object cannot begin and end a change as a physical object, but be something non-physical at some point during the course of this change. If the material cause acts as the persisting substratum in generation and destruction, and every changing object must continuously be a physical object, then it must be the material cause that is responsible for the physical nature of anything undergoing generation or destruction. The formal cause being gained or lost during generation and destruction cannot be responsible for the persisting nature of the thing undergoing these changes, beginning with its persisting physical nature.

The same result is obtained when we consider Aristotle's claim that a material cause possesses the potentiality to be a certain kind of object.[48]

As we saw above, Aristotle emphasizes that not everything can function as a material cause. A material cause possesses the potentiality to be a certain kind of object only if it possesses the features that make it suitable for *receiving* that formal cause; it must be the kind of thing that can be acted upon by some external efficient cause in such a way that the formal cause in question is brought into being in it. In particular, generation presupposes locomotion, because the relevant efficient cause must be able to move to, and come into physical contact with, the material cause of the object to be generated. This contact is possible only if the material causes of both agent and patient are themselves physical, movable bodies. The physical nature of a generated object, then, cannot be grounded in the object's formal cause, because it is possible to bring that object's formal cause into being only if its material cause is already something physical. The physical nature of a generated object's material cause is a necessary condition for the actualization of its formal cause.

Indeed, the above conclusion can be extended to all instances of change. Not only must everything subject to generation and destruction have a physical material cause, but also, in Aristotle's account, there is no change of any kind without matter. As we saw in section 1.1.3, Aristotle argues that locomotion is the first and most fundamental kind of change; every other kind of change, including generation, presupposes locomotion, but not the converse. Furthermore, we have seen that not only must everything capable of locomotion be extended and capable of occupying a place, that is, be a finite physical body, but every such body must also have a physical material cause. The same holds true of everything capable of acting on, or being affected by, perceptible objects. Thus, Aristotle tells us not only that every change must have a persisting substratum, but also what that substratum must be. Since there is no change without locomotion, not only everything movable, but also everything changeable in any way requires a physical substratum, something corporeal and movable in its own right. The latter will be the case only if that substratum is, or is made out of, physical stuff of one kind or another. In the end, something has a substratum for change only because it is made out of matter, and whatever lacks a substratum of change does so precisely because it lacks matter. All and only objects made out of matter are changeable.

Chapter Four

The Material Elements and Prime Matter

As we saw in chapter 3, Aristotle allows for the layering of material causes inasmuch as one and the same thing can both act as a material cause for something else and have a material cause of its own.[1] Bronze, for example, has its own material cause – the elements from which it is made – and can also act as the material cause of a statue. Aristotle thinks that this layering is finite at the lower end in that at the bottom there will be a material cause that does not have a material cause of its own; whatever this material cause is, it will be the first material cause of the object made from it in the sense of the ultimate material cause of that object. In this sense, every perceptible object has a "prime matter," that is, a first material cause from which it is ultimately made.

Chapter 3 also argued that Aristotle's requirements for motion, physical interaction, and generation dictate that the material causes of perceptible objects be something physical in their own right. This chapter argues that when these requirements are applied to the material elements, it follows that they too must be made out of something physical. In other words, the material cause of the elements – traditionally referred to as "prime matter" – must also be something physical in its own right.[2] If this is the case, Aristotle not only has a concept of matter as physical stuff, but also a concept of simple physical stuff, whose nature consists simply in being extended, movable, and divisible. This simple physical stuff is seen most clearly in the four sublunary elements, but is also found in the fifth element, ether.

4.1 The Common Substratum of the Material Elements

Aristotle argues that the four sublunary elements have two kinds of fundamental properties, their natural motions and their tactile properties.[3]

Of the two, the latter are more important,[4] for the tactile properties of the sublunary elements govern their causal interactions with other bodies, including those interactions that lead to the generation and destruction of the sublunary elements themselves; for this reason, Aristotle takes these causal properties to be the most important properties of the sublunary elements. These properties also belong to the sublunary elements in their own right; as we shall see in chapter 5, the natural motions of the elements belong to them only in the context of the ordered physical universe and its limits. Thus, our consideration of the sublunary elements begins with their tactile properties.

Like other perceptible objects, the four sublunary elements are capable of physical interaction with every other extended, movable body. Unlike other perceptible objects, the capacity for physical interaction with other bodies is what defines them. In Aristotle's account, these causal properties are arranged in two pairs of contraries, heat and cold, and fluidity and solidity. Each element possesses one contrary from each pair: fire is solid and hot; air fluid and hot; water fluid and cold; and earth solid and cold.[5] He classifies these contraries as tactile properties, but he is not interested primarily in the way in which these bodies feel to us. Rather, he defines these properties by the way in which these elements behave when they come into physical contact with other corporeal objects.[6] Fluidity and solidity he calls passive capacities, because they characterize how readily a body preserves its shape when struck by another body; heat and cold are active capacities, because they describe the effect that hot and cold bodies have on other bodies when they come into physical contact: heat purifies bodies by causing them to separate into their homogeneous parts; cold causes both heterogeneous and homogeneous bodies to coagulate and condense.

The sublunary elements, then, are defined by the way in which they physically interact with other bodies. As we saw in chapter 1, the capacity for physical interaction requires certain other properties. In particular, the capacity for physical interaction with other bodies can be found only in finite, movable bodies, because only such bodies can affect other bodies.[7] As we saw in chapter 3, physical interaction between bodies presupposes that these bodies have a common substratum; whatever contrary properties interacting bodies may have, these contraries are found in an underlying substratum that is the same in kind in all of the interacting bodies. In fact, this substratum is the only part of the four sublunary elements that is common to them all. As we saw above, their basic properties are taken from two pairs of contraries, heat and cold, and fluidity and solidity. Because contraries alone cannot act on one another, there must be something more to these elements than just the contraries themselves, namely a substratum underlying these contraries.[8] This substratum must be common to the sublunary elements, because all four are capable of physically interacting with each other. While each of the four sublunary elements has one of the basic contraries in common with another element, it is also the case that each element differs from another element with respect to both contraries. Fire and earth are both solid, and air and water both fluid, but fire and water, for example, differ with respect to both contraries: fire is solid and hot, and water fluid and cold. Only the substratum in which these contraries are found is common to all four sublunary elements.

This common substratum must be something physical in its own right. As we saw in chapter 3, the substratum underlying the physical interaction of any two perceptible objects must be a physical body of some kind. This requirement also applies to the elements, indeed, preeminently so, for the fundamental properties of the sublunary elements are the causal powers by which they physically interact with other bodies. Because physical contact between perceptible objects requires a common substratum that is extended, movable, and divisible in its own right, heat, cold, fluidity, and solidity, as Aristotle understands them, can be found only in physical bodies. Since there are no other elements out of which these four are made, their material cause – the substratum in which their defining properties are found – must itself be physical matter.[9] This physical matter must also be found in the fifth element, ether, because, as we saw in section 2.2, it too is capable of physically interacting with the sublunary elements. A physical substratum, then, common to all of the elements, is required to explain their physical interaction.

The same result is reached by considering the natural motions of the four sublunary elements. As we saw in section 1.1.2, the capacity for natural motion presupposes the capacity for motion, and there are certain properties that the material elements must have simply as movable objects, regardless of their natural motions.[10] Aristotle holds, for example, that every instance of locomotion is continuous, that is, it has no gaps through which the moving body does not pass. Only a body continuously extended in three dimensions, he argues, is capable of such motion.[11] Furthermore, anything capable of locomotion must be capable of having a place, and the latter belongs only to physical bodies.[12] In addition, movable bodies are impenetrable inasmuch as it is impossible for two such bodies to occupy the same place.[13] Finally, any continuously extended material body is infinitely divisible, that is, it cannot be composed of unextended parts.[14] Thus, there is no motion unless the object in motion possesses these physical attributes.

It follows that the four sublunary elements must also possess these attributes, including during their reciprocal generation and destruction, for generation and destruction are impossible without locomotion, and locomotion presupposes the existence of extended, divisible bodies that occupy a place.[15] In order to satisfy the requirements for locomotion, the object in motion must be something physical; in the case of generation and destruction, the persisting substratum must satisfy these requirements. These requirements for locomotion also apply to the fifth element, ether. Aristotle makes this connection explicit when he says that the five elements have the "matter of location" (ὕλη τοπική) in common.[16] Thus, extension in three dimensions, infinite divisibility, and occupying a place in relation to other bodies are attributes common to all five elements. Because their substratum is the only part of these elements that is common to them all, these common physical attributes must be found in this substratum.

Like every other perceptible object, then, the elements must be made out of something physical, for subtracting from the elements all of the properties that give them a particular size or shape does not mean that they cease to be extended bodies. All of the properties that define the elements presuppose the existence of an extended body, which these properties qualify or limit. The matter out of which the elements are made, in Aristotle's account, is completely indifferent to any particular shape or size, as well as to the perceptible properties that differentiate the elements from another. As it happens, every perceptible body is qualified in such a way as to be one of the elements or a compound of them. None of these differentiating properties, however, make something to be an extended, movable body in the first place.[17] On the con trary, the elements can acquire these properties only if their substratum is already a physical body in its own right. Thus, the material cause of the elements, traditionally called "prime matter," must be matter. Indeed, the nature of prime matter is the nature of matter *tout court*, because prime matter has whatever is left to matter after all of the differentiating properties of the elements are taken away. This simple matter underlies all of the various elements and is, as a result, the ultimate material cause of all perceptible objects.

4.2 The Generation of the Material Elements

In addition to having natural motions, Aristotle argues that the four sublunary elements are subject to generation and destruction. As with every other change, generation and destruction require a persisting substratum. The substratum persisting through the generation and destruction of the four sublunary elements must be common to them all, because they can all be generated from one another. Sometimes one of the contrary properties persists in the generated element; both air and water are fluid, so when air is generated from water, something hot and fluid comes to be from something cold and fluid. Sometimes both defining contraries are lost; when air is generated from earth, something hot and fluid comes to be from something cold and solid. Thus, the substratum persisting through the generation and destruction of the four sublunary elements must be not only common to them all, but also independent of any of their defining contraries. The only thing that can persist through these changes and act as the substratum that gains and loses these properties is extended, movable matter. Indeed, given that none of the contraries that define the sublunary elements is common to them all, the extension and mobility that are common to them all must belong to their common substratum.

The traditional doctrine of prime matter denies this conclusion by making the contraries responsible for the common corporeal nature of the elements: the substratum underlying the elements is nothing in its own right, or pure potentiality, and so the corporeal nature of the elements must be due to their differentiating contraries.[18] This account fails, however, because the generation and destruction of the sublunary elements requires a substratum that possesses physical characteristics in its own right, for, as is the case with every change, the persisting substratum must already possess in its own right the features that enable it to receive what is gained or lost in the course of that change. The only thing capable of acting as the substratum for the changes in which the defining contraries of the sublunary elements are gained or lost is physical matter. Far from the persisting substratum being dependent upon these contraries for being extended and movable, these contraries depend upon their substratum, because they require an extended, movable substratum before they can come into being at all.[19] The traditional doctrine of prime matter gets the direction of explanation wrong; the substratum is responsible for the common physical characteristics of the elements, not the contraries that differentiate the elements. These contraries cannot be responsible for a physical substratum, because their possibility presupposes the actuality of such a substratum.

In addition, the claim that the contraries impart extension and mobility to the elements suggests that these contraries are essential to their common substratum, that is, that they make it to be what it is. In fact, the opposite is the case: these contraries cannot be essential to the substratum that persists through the generation and destruction of the elements, for none of these contraries belongs to the underlying substratum all of the time; each is lost in at least one of the changes in which the elements are generated or destroyed. Just because the underlying substratum is never without some combination of these contraries, it does not follow that these contraries are essential to it.[20]

Indeed, it is a requirement for all types of change – including generation – that the contraries gained or lost in a change must be accidental to the substratum persisting through that change, for if the substratum is to persist through a change, the properties gained or lost in that change cannot be necessary to it; if they were, the substratum would itself be generated or destroyed in the process of gaining or losing those properties. The differentiating properties of the sublunary elements may be essential to these elements, but their differentiating properties cannot be essential to the substratum that persists through their generation and destruction. As in other types of change, whatever is gained or lost in generation and destruction must be accidental to the substratum persisting through these changes.

The claim that the defining contraries of the sublunary elements are not essential to their common substratum does not imply that there is a kind of matter that exists apart from the material elements. Aristotle repeatedly insists that there are no physical bodies without the properties that make them to be earth, air, fire, water, or ether, or some composite of these elements.[21] There are no elements apart from these five, and every physical body is either one of them or is made of them. The inseparability of prime matter from the elements, however, does not entail that its own nature depends upon the differentiating properties and natural motions of the elements; if anything, these differentiating properties can be understood only in relation to the nature of the substratum in which they are found. Similarly, just because there is no physical matter apart from the material elements, it does not follow that there is no matter in them; the material elements are always more than just physical stuff, but, by virtue of their material cause, they are always at least that. The inseparability of prime matter from the elements, then, does not entail its pure potentiality. On the contrary, if prime matter is to act as a persisting substratum, it must have a nature of its own, independent of the formal causes that it gains or loses.

As a result, when Aristotle argues that essential change among the elements has no perceptible persisting substratum, his point is not that this persisting substratum should be thought of as purely potential.[22] To begin with, at every point in time during which one element is changing into another, it has some perceptible properties, and the perceptible properties of the elements exist only in physical bodies. Furthermore, a change in all perceptible properties does not imply the pure potentiality of the substratum. The substratum itself is potentially perceptible only because it can exist without any of the particular perceptible properties it happens to have at any one time. Nevertheless, it always has some of these perceptible properties, and the latter exist only in matter. The persisting substratum may be only potentially perceptible, but it cannot be only potentially a body; something is potentially perceptible only if it is actually a physical body.[23]

If the above analysis is correct, prime matter cannot be pure potentiality, that is, lacking any determinate nature of its own. As it is traditionally understood, prime matter is not sufficient to act as the substratum required for the changes it is said to underlie. It follows that one cannot argue that Aristotle has no concept of matter because prime matter, the ultimate material cause of all perishable objects, is nothing like what we would think of as matter. On the contrary, for Aristotle matter is necessary for all change, because everything subject to change must, ultimately, be made out of matter. Thus, the proximate material cause of the simplest perceptible objects, the five material elements, must itself possess the basic attributes of matter. In fact, not only must prime matter be matter, but, given that the five elements are made out of it and all other perceptible objects are made out of these five elements, prime matter is also responsible for the corporeal nature of all other perceptible objects. Thus, beyond a concept of matter as physical stuff, Aristotle has a concept of matter as simple physical stuff, and he sees all perceptible objects as ultimately made out of this simple physical stuff. Whatever intrinsic properties all perceptible objects have in common belong to them by virtue of their common substratum, namely the physical stuff in which their differentiating properties are ultimately found.

As we saw in the introduction, the pure potentiality of prime matter is said to be necessary in order to preserve the reality of essential change – generation and destruction – as distinct from accidental change, in particular, alteration. We can now see that this argument is based upon an incorrect understanding of the distinction between these two kinds of change. The defining characteristic of essential change is not that its persisting substratum has no determinate nature of its own. In generation and destruction, as in every other change, the persisting substratum must have a nature of its own, independent of whatever is gained or lost in that change. This requirement means, in turn, that the formal cause that comes to be in generation cannot be responsible for making both the generated object and the substratum persisting through that generation to be the kinds of things they are; the formal cause is essential to the composite object, not to the substratum of that object. As we saw in section 3.3.3, the formal cause gained or lost in any essential change must be accidental to the material substratum underlying that change; this non-persisting formal cause cannot be part of the nature of whatever serves as the persisting substratum. This requirement also applies to prime matter: if it is to persist through an essential change, it cannot be defined in terms of the formal cause that does not persist through that change.[24]

The upshot here is that prime matter cannot be understood simply as the principle of privation in any physical body capable of generation or destruction. With respect to every substratum of change, one has to distinguish between its own nature, on the one hand, and its non-being or privation, on the other.[25] In the case of a block of marble to be sculpted into a statue, for example, one has to distinguish between what it is as marble and what it is as lacking the kind of shape that would make it to be a statue. The privation belonging to a substratum of essential change must be understood in relation to some formal cause or other, and thus is always a restricted kind of non-being. In fact, Aristotle criticizes Plato for making just this mistake: if the material substratum were simply privation, it would seek its own destruction in the process of generation.[26] Because it is only the privation of a formal cause that is lost in generation, the substratum, which is distinct from that privation, can remain as it was before. Whatever acts as a substratum of essential change gains or loses a formal cause in that change, but this relation to the formal cause does not, and cannot, make the substratum that persists through this change to be what it is.[27]

As a result, prime matter possesses a nature of its own. What this nature is can be determined by the characteristics that it must have in order to perform its role in the physical interaction, motion, and generation of the material elements. Given these requirements, physical matter alone is adequate to serve as the proximate material cause and common substratum of the material elements.

Chapter Five

Simple Physical Necessity in the Material Elements

In the previous chapter, we saw that the material cause of the material elements – prime matter – must be something physical in its own right, because only a physical material cause can act as the substratum persisting through the changes that the elements undergo. In this chapter, we turn to the material elements themselves, in particular their distinctive causal capacities. These causal capacities are the basis for the physical necessity that, in Aristotle's account, governs the behaviour of not only the material elements, but all perceptible objects. This chapter considers the material elements; the next chapter looks at the way in which the physical necessity found in the elements also governs the behaviour of the perceptible objects made from them. Subsequent chapters discuss the way in which the higher-level capacities of perceptible objects make use of the physical necessity that governs their lower-level material causes.

5.1 Simple versus Hypothetical Necessity

Aristotle argues that the primary meaning of the term *necessity* (ἀνάγκη) is what cannot be otherwise.[1] He further distinguishes what is necessary simply, or unconditionally, from what is necessary hypothetically, or conditionally.[2] By way of examples of hypothetical necessity he cites the relation between the premises and conclusion of a syllogism, as well as the relation between a composite object and its material cause. The latter is the type of hypothetical necessity relevant to our topic. It holds whenever a material cause of a certain sort is necessary *if* a certain kind of formal cause is to be actualized and the resulting composite object is to function properly.[3] Aristotle's favourite example is that a saw must be made out of something hard, such as iron, if it is to cut wood. In the case of artefacts and generated natural substances, Aristotle thinks the requirement for the right kind of raw materials is only an instance of conditional necessity, because there is no unconditional necessity that the formal cause of these things come to be.[4] As we saw in chapter 2, bringing a formal cause into being typically requires the right kind of material and efficient causes. Thus, even if the appropriate raw materials are available, there is no necessity that the perishable object in question will be generated. If the perishable object exists, the right kind of raw materials must be present, but not the converse.

In addition to this hypothetical necessity, there is a simple, unconditional necessity governing the material elements. This simple necessity comes in two kinds, one teleological and one non-teleological.

5.2 Simple Teleological Necessity in the Elements

The teleological kind of simple necessity is seen in the natural motions of the elements. Every element, Aristotle says, moves by nature to a certain place in the cosmos, or in the case of ether, within a certain place.[5] These natural motions are simply necessary because the elements always move this way and cannot move otherwise.[6] They are also an example of teleological motion in that they are directed to an end. In the case of the sublunary elements, that end is the place to which they naturally move.[7] In the case of the fifth element, ether, its end is the perfection and eternity of its motion, which, Aristotle argues, is caused by the soul in the outermost sphere moving it in this way in imitation of God's perfection.[8] The natural motions of the material elements, then, are an example of simple, teleological necessity.[9]

Aristotle's claim that the natural motions of the elements are simply necessary must, however, be qualified in two ways. First, these natural motions do not belong to the elements in their own right; they are conditional upon these bodies being part of a universe that is physically structured in a certain way. Outside of this ordered universe, they would no longer move in this way.[10] Second, at least in the case of the sublunary elements, their natural motions can be blocked, that is, they can be prevented from moving to their natural place by an external obstacle. The natural motions of the elements, then, seem to be only conditionally necessary in that they depend upon two sorts of external conditions being met. Indeed, some commentators have claimed that in Aristotle's account there is no simple physical necessity at work in nature at all; all natural necessity is conditional, arising from the need of natural substances for the right kind of raw materials.[11]

With respect to the first qualification, it is certainly the case that the natural motions of the elements depend upon a fixed system of places. The elements have natural places in that there are places to which they naturally move.[12] These natural places do not exist in some kind of absolute space, but only as part of a system of places that is ultimately grounded in the outermost sphere of the cosmos.[13] The downward motion of earth, for example, is always towards the centre of the earth; in Aristotle's account, this motion is more accurately described as motion towards the centre of the cosmos, and the latter is where it is because it is the centre of the outermost sphere.[14] The earth is at the centre of the cosmos because that is the place to which earth naturally moves. The same holds for the upward motion of fire, which is motion away from the centre of the cosmos. The natural motions of water and air are to places intermediate between those of earth and fire. Without the outermost sphere, then, there would be no centre point to which earth would naturally move, as well as the other natural places determined in relation to the outermost sphere and its centre.[15]

Despite this dependence upon the overall structure of the universe, Aristotle does not consider the natural motions of the elements to be instances of conditional necessity. The reason is that the features of the universe upon which these motions depend are, in his view, permanent. The natural motions of the elements may depend upon a certain formal cause, but that formal cause is always present in the physical structure of the cosmos. The elements have these natural motions *if* they are part of an ordered cosmos, but the antecedent of this conditional claim holds eternally: the elements are always part of this cosmos, and its order eternally constrains their motions. By contrast, Aristotle considers the necessity that applies to the material causes of perishable objects to be conditional, precisely because it applies only when these objects exist or are being generated.[16] Since perishable individuals do not exist eternally, this kind of necessity is not always at work. More precisely, it holds eternally only at the level of the species: things of a certain kind always require raw materials of a certain kind. At the level of perishable individuals, however, there is no unconditional necessity that their formal cause will be actualized at a particular time and place, or, if it has been, that the resulting composite individual will exist eternally. Only a kind of transitory necessity, therefore, applies to perishable individuals.

At the level of the elements, by contrast, the order of the cosmos that determines their natural place exists eternally.

In the sublunary elements, this simple natural necessity manifests itself in their heaviness and lightness, for heaviness and lightness derive from the natural motions of the sublunary elements: heaviness is simply the tendency to move downward found in elements that are above their natural place; lightness is the tendency to move upward in elements that are below their natural place.[17] The only difference between natural motion and heaviness and lightness is that natural motions are the same regardless of the size of the body and depend solely on the natural place of the element in question; heaviness and lightness, by contrast, vary according to the size of the body, or more precisely, the amount and composition of the elements that it contains.[18] Thus, notoriously, Aristotle claims that heavier bodies made out of earth fall to the ground faster than lighter ones because the latter contain less earth.[19]

There is a second qualification to Aristotle's claim that the natural motions of the elements are unconditionally necessary, namely the fact that the motions of the sublunary elements can be blocked by external obstacles. These external obstacles act in two ways. First, the natural motion of one element can block that of another, through contact between bodies with opposed natural motions; the result is what Aristotle calls "violent motion," namely the movement of a body contrary to its natural motion.[20] Second, the natural motions of the sublunary elements can be blocked by their more fundamental, tactile properties. In some cases, this occurs simply by physical obstruction. In the construction of artefacts, for example, the sublunary elements and bodies made from them can be mixed, tied, glued, or otherwise bound together in a way that prevents them from moving to their natural places. Nature provides many similar examples: Aristotle explains shooting stars, dew, hail, wind, thunderstorms, lightning, hurricanes, and typhoons by appealing to the basic tactile properties of the elements out of which they are made and the physical interactions between them; in many of these natural occurrences, the natural motions of the elements are blocked or overpowered by the natural motions of other bodies, which sometimes result in destructive conflict.[21] He also finds many examples of conflict between the natural motions of the sublunary elements in biological organisms.[22] As a result, Aristotle often qualifies his claims about the natural motions of these elements by saying that they move this way provided nothing external prevents them from doing so.[23]

Despite these external constraints, Aristotle still considers these natural motions to be necessary simply and unconditionally. Here again the reason is that these motions are grounded in something permanent. In addition to the system of natural places within the cosmos, the natural motions of the sublunary elements are grounded in their basic tactile properties, which are among their permanent features.[24] Thus, the tendency to move to their natural place is still in the sublunary elements, even when their motion to that place is blocked. Evidence for this tendency is seen in their behaviour once the external obstacle is removed: they immediately move to their natural place. Moreover, these elements never move naturally in a different direction. In particular, they cannot be "trained" to move in another direction by repeatedly blocking their natural motion or forcing them to move in a different direction.[25] All that stopping their natural motion does is to prevent them from exercising their capacity to move in this way; it does not cause the element to which that capacity belongs to be altered or to lose that capacity. Given the constant generation and destruction of the sublunary elements, there is constant churning among these elements and the objects made from them: when one element is generated from another, it invariably attempts to move to the natural place where it now belongs. In this process, the motions of the elements are frequently in conflict, with the motions of some elements hindering the motions of others. Nevertheless, the natural motions of the sublunary elements remain, for Aristotle, unconditionally necessary, for, in all of these cases, the motions of these elements still follow fixed patterns, as a result of their natural powers and the more general laws of mechanics described in section 2.2 above.

The grounding of simple necessity in the permanent features of perceptible objects is clearest in the case of the natural motions of the planets and stars, for, unlike sublunary objects, there is nothing that can block their motions. Thus, they move always in the same way, without interruption. This constancy also holds for the changes in the sublunary realm caused by them, such as the seasons.[26] Moreover, the natural motions of the heavenly bodies confirm the importance of the material elements for the motions of the bodies made from them. Because the motions of the heavenly bodies are so different from those of sublunary objects, Aristotle argues that the heavenly bodies have to be made out of a different element.[27] The natural motions of the planets and stars would not ensue if their formal causes were added to the sublunary elements; celestial bodies too have to be made out of the right kind of material cause, and the sublunary material elements would not be suitable. Hence, there must be a fifth element, ether.

The natural motions of the five elements, then, are instances of simple, unconditional necessity, because they are grounded in two permanent features of the physical universe: the physical structure of the cosmos and the unchanging capacities of the elements to move themselves. As a result of these two features, the elements can never move naturally in any other way, and when their natural motions are blocked, their behaviour is still due to their permanent properties, either the natural motions of the elements blocking them or their own underlying tactile properties.

5.3 Simple Non-Teleological Necessity in the Elements

The natural, goal-directed motions of the elements are an example of simple, teleological necessity. There are, however, other, more fundamental features of the material elements that display a simple, non-teleological necessity. As we saw in chapter 4, natural motion up or down is not the primary property of the sublunary elements. The properties that define these elements, Aristotle argues, have to be able to explain how they interact with one another, including those interactions that lead to their generation and destruction, and these changes arise only through physical contact.[28] Thus, the tactile properties of these elements are more fundamental than their natural heaviness or lightness, because only their tactile properties can explain their physical interaction with one another.[29] As we saw in section 4.2, all of the tactile properties of the sublunary elements are grounded in two pairs of contrary properties, heat and cold, and fluidity and solidity.[30] These primary properties are defined by Aristotle as the powers that cause bodies to separate, congregate, or change their shape in certain ways when they come into contact with other bodies.[31] These tactile properties also explain the process of mixture, which is crucial to the generation of all other, more complex perishable objects. Indeed, Aristotle takes one of the advantages of his account of the sublunary elements to be that it provides a better account of the mixing of the elements as compared to the theories of his predecessors.[32]

In all of these basic processes, the sublunary elements operate through physical contact between bodies. There is nothing teleological about the way in which the basic tactile properties of the elements operate, because there is nothing about the physical interaction of these properties that is intrinsically good or bad. As we saw section 1.2, Aristotle argues that something happens for the sake of an end or final cause only if it happens for the sake of some good or other; something must be benefited by the change.[33] Thus, it is not the case that the actualization of every potentiality is for the sake of a final cause; in the case of violently caused changes, for example, the things being changed are often harmed. The destruction of a perceptible object might be beneficial for something else, but that introduces a second change, in something different, and the final cause belongs to that second change. The destruction of a perceptible object is not good for that object itself and thus cannot be the final cause of the change in which that object is destroyed.[34] Destruction may be good for the destroyer but is typically not for the thing being destroyed. Every change can be made to be teleological if additional, useful consequences are taken into account; such useful consequences, however, do not make the change that produced them to be intrinsically beneficial to the object being used in this way.

Because final causes presuppose a benefit to the object undergoing the change, the changes produced by the simple physical interaction of the sublunary elements lack a final cause, for, although the basic properties of the sublunary elements are arranged in pairs of opposites, neither opposite in these pairs is intrinsically superior to the other: there is nothing about these opposites that would make the one the perfection or completion of the other. Each is the privation of the other, but neither is better or worse than the other. Cold, for example, is not simply the privation of heat, but a real nature.[35] Thus, the transition from cold to hot is not, in itself, the transition from something bad to something good. Through interacting and mixing with other bodies, the elements are constantly being heated or cooled and made more or less solid; Aristotle does not describe any of these changes, by themselves, as good or bad, beneficial or harmful. They happen simply by necessity.[36] Thus, in and of themselves, none of the effects produced by the basic tactile properties of the elements could act as the final cause of a teleological change.[37]

To be sure, these processes can lead to the generation and destruction of the elements, and Aristotle holds that, in the case of things that have a good, existing is better than not existing.[38] In the case of the sublunary elements, however, generation does not yield a result that is better than what existed before, because the generation of one of these elements always involves the destruction of another. Moreover, the mutual generation and destruction of the sublunary elements always results in something with the same kind of causal capacities. As a result, the generation of the elements is unlike those cases of generation where the result is better by virtue of possessing additional causal capacities; in this respect, Aristotle says, the generation of water from air is not like the construction of a house from bricks.[39] Thus, the generation of an element does not leave the physical world in a more ordered state or its destruction in a less ordered one.

Moreover, many interactions between the elements and physical bodies do not lead to generation or destruction at all; when a warm body is cooled by a cold body, no generation need occur in either body, and neither the cooling of the one body nor the warming of the other need be beneficial to either of them.[40] The same holds for the basic operations by which the elements interact with other bodies: by themselves, there is nothing teleological about physical contact, mixing, acting on, or being affected by another body. All of these operations can benefit or harm the composite bodies in which they take place. These benefits and harms, however, depend upon higher-level properties, not the basic properties of the elements and their physical interactions. At the level of the fundamental properties of the sublunary elements, nature does not act for the sake of an end, because nothing happens here that is good or bad.

Even with respect to the natural, goal-directed motions of the sublunary elements, part of this causal process is again non-teleological. Aristotle insists that the elements are not animate beings; there is no soul and no conscious desire at work in them directing them to their natural places.[41] They cannot, then, consciously know in which direction to move. The direction of their natural motion, instead, is determined by their relative density and rarity. The upper elements are rarer than the lower ones, and as the elements change from fire to air to water to earth, they become progressively denser.[42] Their relative density is connected to their natural motions, because rarer bodies naturally move upward and denser ones downward; in this sense, the upward natural motion of the lighter elements is analogous to the buoyancy of wood in water.[43] The relative density of the elements, in turn, is connected to how hot or cold they are: heat and cold cause things to become rarer and denser, respectively.[44] Because hot things are relatively rare, they are also light and move naturally upward; cold things are comparatively dense and thus move naturally downward.[45]

As we saw in section 5.2, the natural places and motions of the sublunary elements depend upon the outermost sphere limiting the physical universe; without it, the order and motion of the elements would be chaotic. Within the limits set by this sphere, however, the relative density and rarity of the elements provides the mechanism by which they can sort themselves into their appropriate places. The movement of the elements to their natural place is good for them in that it allows them to exercise their natural capacity for this kind of motion, but that capacity belongs to them only insofar as they are part of an ordered cosmos; outside of that cosmos, they lack this capacity. Moreover, part of the mechanism by which they move naturally involves their relative density and rarity, and there is nothing teleological, nothing good or bad for the elements, about density and rarity as such. Thus, they cannot by themselves serve as final causes.

Finally, as we saw in section 2.2, Aristotle describes several other processes in the elements that occur in a way that is simply necessary and non-teleological. The first is the production of heat through friction. Aristotle claims that the heat produced in the sublunary world by the sun arises from the rubbing of the revolving heavenly spheres against the next contiguous element, fire.[46] More precisely, through the sun's contact with the ether that fills the space between it and the outermost fire surrounding the earth, the sun moves a portion of the ether, which in turn affects a contiguous portion of fire. When the ether moved by the sun rubs against this portion of fire, the latter, in turn, rubs against the next contiguous portion of air, and this friction causes these sublunary elements to ignite and produce heat. In confirmation of this explanation, Aristotle appeals to the heat produced in sublunary projectiles by the friction between them and the air through which they move. Furthermore, he argues that the sun alone among celestial bodies produces heat in the sublunary realm because of its proximity to the earth and the speed of its motion; the moon, though nearer, moves too slowly to produce much heat through friction, and the other planets and stars do not move fast enough or are too far away.[47] Again, Aristotle confirms this explanation by appealing to the behaviour of terrestrial projectiles.

Another example of simple, non-teleological necessity is found in the physical interaction between the celestial and terrestrial elements involved in the transmission of light. In his *De anima*, Aristotle says that the perception of colour presupposes the property of transparency in the intermediate bodies transmitting the light, and that this property is common to water, air, and the ether of the heavenly bodies.[48] Moreover, both fire, in the sublunary world, and ether, in the celestial world, produce light in the same way.[49] Light, for Aristotle, is not fire or a corporeal body of any kind, but something transmitted by transparent bodies. Since all acts of perception require either direct contact between the sense organ of the perceiver and the object perceived or contact through intermediate physical bodies, perception of the light from a star requires successive physical contact between the elements that are found between that star and the perceiver.[50] Again, this power of transmitting light and colour is understood to operate in the celestial and terrestrial elements in the same way, independent of the difference in their natural motions.

Finally, in his argument against the Pythagorean theory that the motion of the heavenly spheres produces a musical harmony, inaudible to us, Aristotle again applies his mechanics and non-teleological necessity to celestial bodies. He argues that because the stars move by being carried on spheres made out of ether, their motion is silent, in the same way that the motion of a ship drifting downstream in a river is silent.[51] Sound, he argues, is produced by a body in motion only if it collides with surrounding bodies that are not in motion or are not moving in the same way. The sound resulting from these collisions is proportional to the size of the bodies involved: the larger the bodies in contact, the greater the sound produced. The role of contact in the production of sound is also shown by the fact that loud sounds can affect other bodies at some distance and even set them in motion.

The three examples above involve physical interaction between the elements or bodies made out of them, and in all of these instances the physical interaction involved is explained by Aristotle in a non-teleological way; the causal principles he applies here hold for all bodies, regardless of their natural motion, location, or particular composition. The above examples also confirm the point made in section 2.2 that Aristotle applies his principles of mechanics to both the celestial and terrestrial realms.[52] All of these changes, as well as the operation of the fundamental attributes of the four sublunary elements, their heat and cold, and fluidity and solidity, are governed by non-teleological necessity. In sum, there are two types of simple physical necessity that govern the behaviour of the five material elements, the one teleological, governing their natural motions, and the other non-teleological, governing their physical interaction. Moreover, in the case of at least the four sublunary elements, the non-teleological necessity is causally prior to the teleological necessity, because the former arises from their most fundamental, tactile properties; these tactile properties, in turn, are responsible for the relative rarity and density of these elements, upon which their natural motions depend. The basic causal powers of the elements, then, operate in a non-teleological way, and their teleological motions presuppose these underlying non-teleological capacities.

Chapter Six

Simple Physical Necessity in Objects Made out of the Elements

In the previous chapters, we saw that Aristotle considers individual perceptible objects to be composite beings, consisting of a formal and material cause.[1] This composite character holds not only for physical artefacts such as bronze spheres and iron saws, but also natural substances, everything from plants and animals to flesh, wood, and, ultimately, the material elements themselves. The composite character of perceptible objects has implications for the explanation of their behaviour. In chapter 2 we saw that scientific explanations, in Aristotle's view, must be grounded in the structure and parts of both the agent producing the change and the object on which it is acting. The remaining chapters of this book argue that when identifying the parts to which the causal powers of perceptible objects belong, Aristotle regularly turns to their material causes, for the material causes of perceptible objects possess causal powers of their own that cannot be explained by the formal causes of the objects made from them. Just as the formal cause of a statue does not explain the distinctive features of bronze, so too the formal causes of naturally occurring objects do not explain the distinctive features of the materials from which these objects are made. In fact, far from explaining the causal capacities of their material causes, the formal causes of perceptible objects depend upon them. The upshot is that some aspects of the regular behaviour of perceptible objects must be explained by the material causes from which they are made. In particular, as this chapter argues, the physical necessity that operates in the elements also governs the behaviour of all of the perceptible objects made from them.[2]

6.1 Simple Teleological Necessity in Perceptible Objects

As we saw in section 5.2, Aristotle says that all of the elements move naturally to, or within, a certain place in the cosmos.[3] These movements are necessary in that they always follow the same pattern and, given the fixed physical structure of the universe, cannot take place in any other way. These natural motions, however, are also seen in everything made from the elements. All perishable objects, for example, move naturally to the natural place of the sublunary element from which they are predominantly made,[4] for the natural tendency to move to these places belongs primarily to the sublunary elements; it belongs to other perishable objects because they are made out of these elements.[5] The same holds for the heaviness or lightness of perishable objects, which is again due to the heaviness or lightness of the elements from which they are made.

In the case of the fifth element, ether, because the bodies made from it are eternal, we never observe it apart from the heavenly bodies. Thus, it is not possible to compare its motion apart from these bodies with its motion when part of one of them. Still, as we saw in section 5.2, the heavenly bodies cannot be made from just anything; their material cause matters. In particular, the material element from which these bodies are made cannot be one of the sublunary elements, because it has a natural, circular motion of its own, which is different in kind from the natural, rectilinear motion of the sublunary elements.[6] Here Aristotle argues that because the elements can have only one natural motion, and circular motion is as basic as rectilinear motion, the different natural motion of the planets and stars requires that they be made out of a different material element.[7] Indeed, not only is the natural motion of ether different from that of the sublunary elements, it is also effortless: because the planets and stars have no contrary motions to overcome, their motions occur without any resistance.[8] Here again the motions of perishable objects are different: because perishable objects always contain some admixture of multiple elements, their motions always involve some resistance from their materials in addition to the resistance of the plenum through which they are moving.

Like other perceptible objects, biological organisms move to a natural place because of the elements from which they are made. In particular, all plants and animals have a natural tendency to move downward.[9] When at rest, all plants and animals are found on the earth or in water, because earth and water are the elements from which they are predominantly made.[10] If one argues that this regular pattern of motion is not due to their material cause, it must be due to their formal cause, their soul. In this case, however, one and the same movement, the downward natural motion of plants and animals, will be explained in many different ways, namely by the different kinds of soul possessed by the different kinds of biological organism. Aristotle, however, treats motion to a natural place as a single type of occurrence, to be explained in the same way in all cases, namely by the natural motions of the material elements. Thus, his explanation is independent of the fact that plants and animals are animate beings; he explains the downward natural motion of inanimate bodies made from earth and water in the same way. It follows that Aristotle's explanation of downward natural motion in biological organisms is independent of their formal cause, their soul.[11] Nor can one argue this motion is an accidental property of biological organisms; plants and animals always move this way, and anything that happens always or for the most part is not accidental.[12]

Another indication that Aristotle sees the natural motions of perceptible objects to their natural place as due to the material elements from which these objects are made is that these motions cannot be set aside by the formal causes that are added to the elements to produce more complex objects. On the contrary, in Aristotle's account, there is often a conflict between the several natural motions of perceptible objects in that the ones due to their material cause can work against the ones due to their formal cause.[13] If all natural motions were due to the formal causes of perceptible objects, there should be no such conflict. In fact, in Aristotle's account, the distinctive kinds of motion and rest that belong to the different species of natural substance are subordinate to the natural motions of the elements. The specific behaviour of perceptible objects must take place within the confines of what is physically possible, which is determined, ultimately, by the elements.

The conflict between the natural motions of the elements and those of the more complex objects made from them arises because in Aristotle's account the material elements act as material causes twice over. In the first instance, the material elements are the material cause of the cosmos as a whole, inasmuch as each element naturally occupies a particular region within the cosmos; taken together, these regions exhaust the extent of the physical universe.[14] In effect, the perceptible universe is made out of the elements. In addition, within these different regions, the material elements are the material causes out of which all other perceptible objects are ultimately made, whether they be plants, animals, metals, houses, clouds, or stars. The role of the material elements as material causes of the cosmos takes precedence over their role as material causes of particular perceptible objects: the formal causes of the latter cannot set aside the natural motions of the elements from which these objects are made. As a result, everything else that composite perceptible objects do as members of a particular species is constrained by the tendency of the elements to move to or within their place in the cosmos.

6.2 Simple Non-Teleological Necessity in Perceptible Objects

The behaviour of all perceptible objects, then, is subject to the natural motions of the elements. As we saw in section 5.2, however, the natural motions of the elements are themselves subject to two other constraints: the elements can be moved violently, contrary to their natural motion, and the natural motions of the elements can be obstructed by their more fundamental, tactile properties. These two constraints also apply to the perceptible objects made from the elements: these composite objects can be made to move violently, contrary to their natural motion, through contact with a body with an opposed natural motion, and a heavy body can be prevented from falling downward if it comes into contact with another body rigid and solid enough to stop the heavy body from passing through it.[15] Thus, the same constraints on the natural motions of the elements are found at the level of the complex perceptible objects made from them.

The spatial cohesion of perceptible objects is a good example of the tactile properties of the elements working against their natural motions, for the natural tendency of the material elements to move to separate places in the cosmos raises a problem for any physical body that is composed of more than one element. Given the naturally divergent motions of the elements, something is required to hold materially heterogeneous objects together.[16] In the case of living beings, their formal cause has a role here, but the problem of accounting for the spatial cohesion of physical bodies is not unique to living things; it holds for inanimate bodies as well, so the soul cannot be the general solution to this difficulty.

Again, part of the answer lies with the material elements. Aristotle argues that all four sublunary elements must be found in those homogeneous mixtures whose natural place is close to the earth,[17] for, if these mixtures are to have a spatially well-defined body – as opposed to being just a heap of dust – they must contain a certain combination of these elements. In particular, they must contain water because it is the most fluid, or easily shaped, of the elements, and this property is necessary for the cohesion of material compounds: without it, anything made out of earth would fall apart. More precisely, both earth and water are required, because each serves as a kind of glue to the other, thereby holding together the body made out of them.[18] In the first instance, this requirement applies to homogeneous mixtures, but because all biological organisms are made out of these mixtures, they too must contain this combination of elements; otherwise, they too would fall apart. In general, to have a spatially cohesive and well-defined body, all perishable objects, including biological organisms, must be made from a certain combination of the sublunary elements.

A similar dependence on the elements is found in Aristotle's account of solidification and liquefaction. All perishable bodies with definite spatial limits are solid, in some measure, and have some degree of hardness or softness.[19] Their degree of hardness or softness, together with related properties such as pliability, ductility, and compressibility, result from the processes of solidification or congealing, on the one hand, and liquefaction or melting, on the other.[20] Solidification and liquefaction, in turn, arise from the tactile powers of the sublunary elements, their heat, cold, fluidity, and solidity.[21] While these powers are not found exclusively in the sublunary elements, they are the defining properties of these elements and belong primarily to them.[22] As a result, when Aristotle explains solidification and liquefaction in perishable objects, he does so by appealing to the sublunary elements from which they are made.[23] In particular, the reaction of perceptible objects to heat and cold depends upon the elements in them. Because both heat and cold can cause solidification and liquefaction, there is no simple pairing of the two active powers, heat and cold, with these two processes. Which effect heat and cold produce in a physical body is determined by the combination of elements from which that body is made: bodies made predominantly of earth are solidified by heat and liquefied by cold, as in the baking of clay and the dissolution of certain foods; bodies made predominantly of water are solidified by cold and liquefied by heat, as in the freezing and melting of liquids.[24]

Given his pattern of interaction, Aristotle now has a test for determining the material composition of homogeneous mixtures. Because metals such as gold, silver, and bronze, for example, are solidified by cold and liquefied by heat, he concludes that water predominates in them.[25] In the same way, he determines what combination of elements is present in compounds as diverse as olive oil, lead, semen, wine, urine, whey, honey, milk, blood, and wood.[26] He applies this method of analysis to plants and animals as well: again, the occurrence of solidification and liquefaction caused by heating and cooling is explained by means of the elements out of which these organisms are made.[27] It also does not matter whether these processes take place in naturally occurring objects or human artefacts such as bricks and earthenware vessels.[28] Solidification and liquefaction under the influence of heat and cold are uniformly explained by the elements from which perceptible objects are made, and not by the specific nature of the object in which these processes occur.

The powers of the elements are also crucial to concoction (πέψις), another process found in many different kinds of perceptible object. Concoction arises from the application of heat to moist raw materials.[29] In this process, the raw materials are "mastered" by the active elemental powers in such a way that the result is something denser, warmer, and more uniform than the original materials.[30] Concoction is widespread, occurring in processes as diverse as digestion in living organisms, the fermentation of wine, and the curing of clay. Here again, Aristotle does not distinguish between naturally occurring and artificially caused instances: his favourite examples of concoction are the boiling and roasting used in the cooking of food.[31] Another example is the solidification of the bones of animals, which takes place by the same process of baking that is used to make earthenware vessels.[32] Examples of concoction, then, are found in biological organisms, human artefacts, and inanimate natural substances. In all of these cases, heat is applied to certain raw materials, and, if the right combination of elements is present, concoction takes place.

Concoction, in turn, is crucial to biological organisms, because it is necessary for their principal biological functions. Perhaps most important, it is the process by which their food is digested.[33] In the case of animals, heat from their body is applied to the food in their gut, and the resulting concoction produces nourishment and separates out what is to be excreted. In the case of plants, the concoction of their food takes place in the earth before being ingested and uses the warmth of the soil. In both cases, concoction produces a nutritive fluid, which can then be distributed throughout the organism to nourish its various parts. From the nutritive fluid, homogeneous compounds such as flesh, bone, tissue, and sinew are produced, from which, in turn, the functional parts of biological organisms are made; here again, concoction is required.[34] By the end of this process of multiple concoctions, the food of a living organism has been assimilated to that organism.[35]

Concoction is also required in biological organisms that reproduce sexually in order to produce their sperm from their nutritive fluid,[36] for the semen and menstrual fluid from which blooded animals are generated, and their analogues in bloodless animate beings, are themselves ultimately concocted out of the food ingested by these organisms.[37] In effect, the capacity for reproduction and growth in biological organisms is the result of their capacity for self-nourishment.[38] Concoction is the process by which all of this takes place. Thus, concoction is necessary for the generation, growth, and nourishment of biological organisms, and if they lose the heat they require to perform it, they die. The elements and their basic powers, in turn, are fundamental to concoction. Indeed, Aristotle goes so far as to say that the elemental powers of heat and cold that control concoction also control life and death, sleep and waking, growth and aging, and disease and health.[39]

In addition to these general biological processes, Aristotle also uses the elements to explain some of the particular features of individual biological organisms or their species. He uses the elements, for example, to explain the temperament of certain animals. Because of the elements from which their blood is composed, some animals are timorous, others quick to anger.[40] Blood has this effect on the temperament of blooded animals because it is the matter from which their whole body is made, for blood is not a part of their body; on the contrary, the heart and blood vessels act as a jar containing the blood.[41] Thus, blood itself is not alive.[42] To support his claim that this influence on the character of animals is due to the composition of their nutritive fluid, and not to the fact that it is blood, Aristotle says that the same holds for the analogous fluid in bloodless animals; thus, bees and other similar animals are more intelligent than even some blooded animals because their nutritive fluid is thin and cold.[43] Thus, the nutritive fluid in many animals, whether blood or something else, is responsible for some of the character traits of the animals made from it due to its particular composition from the elements.[44] Indeed, all of the fluids in the bodies of biological organisms are subject to certain constraints, given that they are predominantly made from water. Just as rain is formed from vapour that cools in the air, condenses into water, and falls to earth, so too fluids such as phlegm originate in the brain when it is unusually cold, leading to vapours from the body condensing there and flowing downward.[45] Again, a process inside a biological organism is explained by appealing to the behaviour of the elements apart from that organism.

In addition, Aristotle devotes most of Book V of his *De generatione animalium* to features or changes in biological organisms that are due to the material elements or the inanimate homogeneous mixtures made from the elements. He explains the difference in eye colour in animals, for example, by differences in the depth and transparency of the watery fluid in their eyes; this explanation he confirms by pointing out that seawater also varies in colour, again as the result of differences in the depth and transparency of the water.[46] This claim builds on his general discussion of transparency and the perception of colour, which we discussed in section 5.3. Given his general requirement for contact between the perceiver and the object perceived, whether directly or through intermediate bodies, Aristotle argues that the perception of colour presupposes the property of transparency in the bodies between the perceiver and the coloured object; since water, air, and the ether of the heavenly bodies can all transmit colour, they must all possess this property.[47] In particular, all three of these elements instantiate the principle that variations in the intermediate transparent body will affect the transmission of light and, thereby, the colour perceived; the process going on in the eye is no different.

Similarly, Aristotle claims that changes in the pitch of animals' voices are brought about by changes in the tautness of their vocal chords or in the temperature of the air passing over them.[48] By way of confirmation, he again appeals to the physical properties of inanimate objects: the same thing happens, he says, in the different sounds produced by the looser and tighter strings of a loom and by the varying temperature of the air blown into musical pipes. Here he applies the same principles that he uses to argue against the Pythagorean theory of the musical harmony of the heavenly spheres, discussed in section 5.3. As we saw, his view is that sound is produced by a body in motion only if that body strikes surrounding bodies that are not in motion or are not moving in the same way.[49] Thus, variations in the motion and composition in the colliding bodies changes the sound they make.

In all of the above, certain properties and processes in biological organisms are explained by the properties of the elements from which these organisms are made. Some of these features are accidental to biological organisms, such as their eye colour; others, such as the failing of their eyesight in old age, are not.[50] In all of these processes, the direction of explanation is from the elements to the things made out of them, not the converse. Thus, the implication is that some of the basic processes that take place in perceptible objects are due to the causal powers of the elements. These causal powers operate by simple necessity, independent of the formal cause of the object in which they are found.[51] Thus, the simple physical necessity that governs the elements also governs the processes that take place in the perceptible objects made from them, including living things.[52]

6.3 Generation of Homogeneous Mixtures

As we saw in the introduction, one argument frequently made against the existence of any unconditional physical necessity operating in all perceptible objects is based upon the claim that the material elements could impose such a necessity on the objects generated from them only if they persisted in these objects, and they do not. Rather, it is claimed, the raw materials from which perceptible objects are generated are so transformed in the process of generation that they no longer persist in the resulting objects. The upshot is that any simple physical necessity operating in the material elements cannot determine the behaviour of the objects generated from them. Where the elements do persist in the objects made from them, say, in physical artefacts, no generation has taken place.

This claim, however, is based upon an incorrect understanding of generation. As we saw in section 3.3.3, generation, like every other change, requires a persisting substratum; in order to persist through a change, the substratum in question must have a nature of its own, independent of what is gained or lost in that change. Stated differently, what is gained or lost in a change must be accidental to the substratum that persists through that change; what does not persist cannot be part of what does. In the case of generation, the persisting substratum is the material cause.[53] Thus, generation is not the generation of the material or formal cause, but of the composite of them.[54] In effect, an object is generated by bringing together a formal and a material cause, and both causes persist in the generated object. In fact, since the formal cause of a perceptible object exists only in a material cause of some kind, whatever acts as a material cause must continue to exist if the formal cause being actualized in it is to persist as well.[55]

The above account of generation appears to be inconsistent with what Aristotle says about the generation of homogeneous mixtures, and if it does not hold in the generation of these mixtures, it does not hold in the generation of all of the things made out of homogeneous mixtures, such as biological organisms. The central text here is Aristotle's claim that earth, water, air, and fire persist only potentially when made into homogeneous mixtures such as gold, wine, bone marrow, flesh, and blood.[56] This statement has been taken to mean that the elements do not persist in the homogeneous mixtures generated from them, and that the essential properties of the elements that do persist, such as natural motion to a certain place, now belong to these mixtures because of their formal cause, not their material cause.[57] In sum, because the material elements no longer exist in the perceptible objects generated from them, they cannot be responsible for the behaviour of these objects.

This claim, however, does not follow from what Aristotle says about the generation of homogeneous mixtures. Ordinarily, the claim that something no longer actually exists means that it has been destroyed and reduced to its material cause. Aristotle, however, is careful to indicate that this is not what he means by the potential existence of the elements in homogeneous mixtures. In fact, he explicitly states that the elements are not destroyed when they combine to form these mixtures, and that the change they undergo is alteration, not destruction.[58] More specifically, the elements undergo what he calls "unification."[59] He claims that the elements do not exist actually in homogeneous mixtures, because he wants to preserve the distinction between a true mixture (μῖξις) of the elements, on the one hand, and their mere spatial aggregation (σύνθεσις), on the other. Homogeneous mixtures, in his view, are homogeneous throughout; every part of them, no matter how small, contains all of the elements out of which they were made.[60] They are not mere spatial aggregates, whose parts are different in kind from one another. The mixing of the elements, then, does not destroy them, but they also no longer exist as separate bodies.

How, then, is the potential existence of the elements in homogeneous mixtures to be understood? The elements persist only potentially here because they no longer have the particular degree of heat, cold, solidity, and fluidity that defines each of them. As we saw in section 4.1, each of the four sublunary elements has a unique pair of the four basic tactile properties: heat, cold, fluidity, and solidity.[61] Unlike other bodies, the sublunary elements have these properties to an extreme degree. Fire, for example, is not just hot; it is extremely hot. When these elements are mixed together to form homogeneous mixtures, the result is something with properties between the extremes possessed by these elements alone. These intermediate states are understood by Aristotle to exist in homogeneous mixtures only as combinations of the more basic properties possessed by the sublunary elements,[62] for homogeneous mixtures do not arise simply through the dilution of one element by another. If such were the case, the addition of any two elements, in any proportion, would result in some kind of homogeneous mixture. Instead, the elements being mixed must be roughly equal in quantity.[63] Moreover, homogeneous mixtures are more complex than the original elements from which they were made; they are not just equally simple stuffs that exist between the elements.[64]

Thus, Aristotle's claim that the elements do not exist actually in homogeneous mixtures arises from his view that the elements persist in these mixtures only in combination with one another, and not as spatially discrete parts. Still, the elements do persist in homogeneous mixtures. In particular, both the corporeal matter out of which the elements are made and their defining powers, their heat, cold, fluidity and solidity, continue to exist in the resulting compound, although now in a certain combination. It is not so much that the defining powers of the four sublunary elements are lost, as that they are combined with one another to form a kind of physical stuff that is more complex than the elements alone. Homogeneous mixtures result from the combination of the essential powers of the elements, not the loss of these powers and their replacement by something else.[65]

Thus, the change that the elements undergo in forming homogeneous mixtures falls between two other kinds of change. On the one hand, there is the process in which the elements are mixed together to form heterogeneous aggregates analogous to sand and gravel; here the elements remain largely unaffected. On the other, there is the process by which the elements come to be from one another, where the generation of one element always involves the destruction of another. Unlike these two processes, the formation of homogeneous mixtures involves neither the destruction of the elements nor their preservation in an unaltered state. They must, then, be preserved in an altered state. Stated differently, the formation of these mixtures follows what one might call the "construction model" of generation in which the pre-existing raw materials persist and acquire some additional structure or capacity. The potentiality of the pre-existing raw materials to become a certain kind of object follows from their capacity to acquire additional attributes, not from their being left behind in the process of generation.

Another problem with saying that the elements are destroyed when homogeneous mixtures are generated from them is that Aristotle continues to explain a wide range of properties that occur in these mixtures on the basis of the elements from which they were generated. As we saw above in sections 6.1 and 6.2, he uses the powers of the elements to explain, among others, the natural motion, spatial cohesion, and degree of solidity in homogeneous mixtures. These processes are not just involved in the generation of homogeneous mixtures, but also in their behaviour once they have been generated. The properties that cause these processes belong, in the first instance, to the elements, and not to the mixtures made from them. Thus, when Aristotle explains the behaviour of homogeneous mixtures by appealing to the elements out of which they are made, he is not just appealing to their pre-existing raw materials. He is explaining how they behave here and now on the basis of what is present in them here and now.[66]

Aristotle does say that the raw materials are "mastered" by the formal cause in the process of concoction by which homogeneous mixtures are generated.[67] As we saw in the previous section, this process involves the action of heat on the appropriate raw materials.[68] If the amount of heat is sufficient and the raw materials are mastered, concoction leads to the generation of a homogeneous mixture; if the heat is insufficient and the raw materials are not mastered, an incomplete object is generated or the process fails altogether. The mastering of the raw materials in this process, however, cannot be their destruction. Concoction presupposes the powers of the elements and results from their mixture, not their destruction. As a result, even after homogeneous mixtures have been generated, Aristotle continues to explain many of their properties by means of the distinctive powers of the elements from which they were generated. In other words, the powers of the elements are at work not just during the generation of homogeneous mixtures; they continue to work in *existing* homogeneous mixtures after generation.[69]

The same holds for the more complex objects that are made from homogeneous mixtures. In particular, the influence of the material elements on the behaviour of biological organisms is pervasive. All of their biological functions presuppose non-biological physical changes and characteristics, everything from the natural motion of biological organisms to a certain place in the cosmos, to the spatial cohesion of their limbs and body parts. Perhaps most important, biological organisms are dependent on the material elements in the course of their generation and growth, for, in Aristotle's account, the capacity for reproduction and growth in biological organisms is the result of their capacity for self-nourishment, and the elements and their basic powers are fundamental to this process. Thus, even in the most fundamental biological processes, living organisms remain dependent upon the material elements and their causal powers, and much that goes on here takes place independent of their formal cause, their soul.

If the elements no longer existed in the perceptible objects generated from them, the most one could say is that the natural motions and changes that elsewhere are due to the elements are caused in perceptible objects by the elements from which they *were* made; the downward motion of plants and animals, for example, would be due to the material elements from which they *were* generated, but that no longer exist in them. In effect, this is to argue that it is possible for the distinctive powers of the elements to exist apart from the elements themselves. This view, however, is contrary to Aristotle's account of efficient causality. As we saw in chapter 5, for example, Aristotle explains the natural motions of the elements by the tactile properties that they have at the time of their motion, not ones that existed earlier in the elements from which they were generated. Nor can the powers by which these motions are produced simply be detached from the elements to which they essentially belong, and then be transferred to other perceptible objects; these powers require very specific physical characteristics in the perceptible object to which they belong. Thus, it is not enough to say that perceptible objects have their natural upward and downward motions because of the elements from which they *were* made. If these natural motions persist, then so must the elements to which they belong, even if the elements persist only in an altered form. The same holds for the other distinctive powers of the elements at work in perceptible objects, where, again, he typically explains the presence of these powers by the presence of the elements to which they belong. Thus, it is not just the generation of homogeneous mixtures that is directly tied to the elements.[70] Even after their generation, Aristotle repeatedly detects the presence of the elements themselves, and not just their causal powers, in homogeneous mixtures and the more complex perceptible objects made from them.

The upshot is that one cannot both attribute to homogeneous mixtures certain natural motions and properties that belong primarily to the sublunary elements and, at the same time, deny the presence of these elements in these mixtures. Because the processes of heating, cooling, solidification, and liquefaction occur in the same way in all sorts of different physical objects, they are typically not sufficient to account for the particular nature of the perceptible object in which they occur. In addition, some kind of formal cause is required, even if that is no more than the proper proportion of the elements to be mixed.[71] The powers of the sublunary elements, however, are necessary, and part of what is going on in these changes is due simply to these elemental powers. Having the appropriate formal cause does not eliminate the need for the right kind of raw materials and the powers those materials bring.

In sum, part of the regular behaviour of homogeneous mixtures is explained by looking "downward" to their material cause. Regardless of the formal cause gained in the generation of these mixtures, they have certain non-accidental capacities because of the elements from which they are made. The non-uniform parts of biological organisms are made, in turn, from homogeneous mixtures, and from these non-uniform parts, biological organisms. The result is that the distinctive causal powers of the elements continue to work in all of the things made out of them, and they do so independent of the formal cause of these more complex objects. If the distinctive causal powers of the elements persist, so too must the elements. In the generation of homogeneous mixtures, for example, the elements are not destroyed, Aristotle argues, because their powers are preserved.[72] Such a claim is hardly surprising, because nowhere else does Aristotle allow the defining powers of one kind of object to be transferred to another kind of object. The elements exist in homogeneous mixtures only in combination with one another. Existing in combination with something else, however, does not entail not existing.

Chapter Seven

The Dual Nature of Perceptible Objects

In chapters 5 and 6, we considered Aristotle's account of the two basic kinds of necessity that govern perceptible objects: a simple, unconditional necessity that operates in the material elements and everything made out of them, and a hypothetical or conditional necessity that holds between composite objects and the raw materials from which they are made. In this chapter, we consider Aristotle's claim that perceptible objects have a dual nature, one due to their formal cause and another to their material cause.[1] This dual nature is clearest in the case of physical artefacts such as houses and drinking cups, but also holds for all naturally occurring objects, everything from marble and wood to plants and animals. Thus, it is found in all perceptible objects.

The dual nature of perceptible objects is fundamental to Aristotle's natural philosophy, for several reasons. First, it underlies the compatibility of the two kinds of necessity discussed above; as we shall see, perceptible objects can be subject to these two very different types of necessity because they have two quite different natures. Second, it means that understanding perceptible objects requires us to consider both of their natures, where the term *nature* is understood to refer to properties that are both necessary to an object and part of what makes it to be what it is. One might call the nature of something in this sense its essence, but the traditional account of Aristotle's material cause holds that the essence of naturally occurring perceptible objects is exhausted by their formal cause. As we shall see, Aristotle's analysis of perceptible objects – including natural substances – leads to exactly the opposite conclusion, namely that their nature in the above sense cannot be captured just by their formal cause. In effect, the dual nature of perceptible objects means that they have two sets of properties that are both necessary to them and part of what makes them to be what they are. This dual nature also has two more important consequences. First, the difference between naturally occurring perceptible objects and physical artefacts cannot be that the former have one nature and the latter two; another explanation for the difference between them will have to found. Second, the unity of natural substances will have to be understood in such a way as to accommodate their dual nature; their unity cannot be grounded in their simplicity. These four topics are considered below.

7.1 The Dual Nature of Physical Artefacts

As we saw in section 5.1, Aristotle argues that there is a relation of hypothetical necessity between composite objects and their material cause. This relation holds whenever a material cause of a certain sort is required if a certain kind of formal cause is to be actualized and the resulting object is to function properly.[2] Two points about this kind of hypothetical necessity should be noted. First, it is an instance of a means-end relation inasmuch as the composite object uses some of the permanent features of its raw materials in order to perform its own functions.[3] In the case of physical artefacts, this dependence on their raw materials is clear: wood, bricks, silver, and iron possess certain properties in their own right, and the houses, drinking cups, and tools made from them need these properties in order to perform some additional function. Second, the requirement for certain properties in the raw materials does not explain the occurrence of these properties in the raw materials. The saw depends on the hardness of the metal from which it is made in order to cut wood, but the metal does not depend on the saw for its hardness. The hardness of the metal is part of the explanation of the saw's capacity to cut wood, not the converse.

There is, then, an asymmetrical relation between a physical artefact and its raw materials: the artefact depends on its raw materials for certain properties, but the raw materials possess these properties in their own right, independent of the artefact made from them.[4] Furthermore, this kind of hypothetical necessity does not belong to the raw materials intrinsically; it holds of them only in relation to the physical artefact made from them. Like the role of the material cause in general, the need for certain raw materials is extrinsic to those materials, holding of them only in relation to the end they serve; this extrinsic, relational property, however, presupposes certain intrinsic properties in the raw materials, the ones needed by the artefact.

Thus, the need for certain raw materials presupposes that those materials contribute something of their own to the physical artefact made from them: the required properties are present in the artefact because they are found, in the first instance, in its material cause. The material cause is only potentially the artefact that can be made from it, but this potentiality is not just due to the actuality that the material cause lacks; it implies as well that the material cause already possesses certain properties in its own right. Just what these required properties are will vary from artefact to artefact. Still, these required material properties belong to completed artefacts only derivatively, because of the raw materials from which they are made.

Thus, the requirement for a certain material cause also points to a deficiency in the corresponding formal cause: physical artefacts require a certain material cause, because their formal cause lacks something that their material cause has to supply. It follows that the formal cause of a physical artefact is not responsible for all of the properties that that kind of artefact requires; if the nature of an object consists in the set of properties required to make it to be a certain kind of thing, part of an artefact's nature is due to its material cause. As a result, the art that knows how to use an artefact is different from the one that produces it: the former, the "ruling" art, as Aristotle calls it, knows the formal cause and function of the artefact; the latter, the producing art, must know its material cause as well.[5] In order to produce an artefact, an artisan needs to know more than just its distinctive properties; the artisan also needs to know the prior, independent features of the available raw materials. The ruling art specifies the properties that are required in the raw materials, but only the producing art knows which raw materials have these properties and how to bring the formal cause into being in those materials. If there were nothing more to know about a physical artefact beyond its formal cause – nothing, that is, to know about the material cause in its own right – the ruling art and the producing art would be the same. The producing art may be subordinate to the ruling art because the latter knows the proper use of the artefact, but the producing art knows something about the artefact that the ruling art does not, namely the nature of its material cause. This material cause has a distinct nature of its own, and without it, the artefact as a whole cannot function. In sum, physical artefacts have a dual nature inasmuch as their permanent, necessary attributes are drawn from both their formal and their material cause.

7.2 The Many Natures of Natural Substances

It is not just physical artefacts, however, that have this dual nature; it is also found in natural substances, where "natural substance" is understood to refer to naturally occurring perceptible objects, ones not made by human beings. In Book II of his *Physics*, Aristotle says that natural substances have two natures, one grounded in their formal cause and another in their material cause.[6] In making this claim, Aristotle is considering the proximate material causes of natural substances, the material causes from which they are immediately made.[7] As we saw in section 3.1, however, Aristotle allows proximate material causes to have a material cause of their own. Thus, some natural substances have several layers of material causes. As we saw in chapters 5 and 6, the physical raw materials that act as the material causes of natural substances have causal powers of their own. Because of the presence of these different kinds of raw materials, natural substances move, or are moved, in several ways; they have motions that are natural to them because of their material cause. Because the nature of a natural substance is captured by its capacity to move itself or to be moved in certain ways, the number of natures in natural substances increases with the addition of such capacities for self-motion. The result is that natural substances have as many natures as they have discrete capacities for moving themselves or for being moved.

The important point here is that these causal capacities belong to natural substances by virtue of both their material and formal causes, and the latter remain distinct, for these causal capacities do not collapse into just one nature because they occur in an ordered sequence. The sequence of material causes and their causal capacities is ordered inasmuch as the higher-level ones presuppose the lower-level ones.[8] The higher-level capacities, however, do not explain the lower-level ones. The relative density and rarity of the sublunary elements, for example, causes them to move naturally to a certain region of the cosmos, which, in turn, is the basis for their heaviness or lightness. The biological structure of plants and animals presupposes these natural motions – in particular, the heaviness of earth and water – but does not explain them. To be sure, biological organisms have natural motions of their own, over and above the natural motions of the elements.[9] These distinctive biological motions, however, do not explain the natural motions of the elements, not even when the latter take place in biological organisms. The independence of the natural motions of the elements is also shown by the way in which they can come into conflict with the natural motion of the composite object made from them.[10] The natural motion of most birds is to fly, but in so doing they must constantly work against the natural tendency of their body to fall to the ground. Ultimately, this conflict leads to the decay and destruction of these organisms. Such a conflict, however, presupposes that there is more than one natural tendency to move at work in them.

Similarly, the property of pliability belongs, in the first instance, to homogeneous mixtures; other perceptible objects have the degree of pliability that they do because of the kind of homogeneous mixture from which they are made. To explain pliability by appealing to anything other than a homogeneous mixture is an explanatory category mistake: saw blades and the bones of animals do not possess their particular degree of pliability because of the artefact or biological organism to which they belong. The same sequence of natural capacities is seen in the heavenly bodies: they have both their celestial motion and the changes that belong to them because of the ether from which they are made, for example, the way in which ether transmits the light emitted by these bodies.[11]

The upshot is that all natural substances have more than one natural motion. The causal capacities of the material elements are not sufficient to explain the behaviour of the natural substances made from them, but they are necessary to the latter and exhibit characteristics of their own, independent of the substances made from them. These elemental causal capacities also set the limits for what is physically possible, because they are prior to all of the other causal powers in perceptible objects. Thus, natural substances are composite entities not just in the way in which every changeable object is something composite, that is, one in number, but two in kind. Natural substances are also composite entities in a more radical sense, namely inasmuch as they have many natures.

These many natures become clear when Aristotle considers the question of what nature is, or more precisely, what exists by nature. His most systematic discussion of this topic is found at the beginning of Book II of his *Physics*. Here he distinguishes six different ways in which we can think about nature and what exists by nature. He begins with the ways in which nature can be understood as a class of things, which he gets at by dividing the class of perceptible objects into natural and non-natural objects.[12] Interestingly, he makes this division twice, using different criteria. The first time, he divides the class of perceptible objects into those made by human beings and those not so made; here nature consists in the class of perceptible objects that exist without human intervention, as opposed to physical human artefacts. The second time, he divides the class of perceptible objects into those that can move or change themselves and those that cannot; here the class of things that exist by nature consists of perceptible objects that are capable of self-caused motion or change.[13]

Aristotle takes these two ways of distinguishing natural from non-natural objects to be largely co-extensive. There are some exceptions: bird's nests and spider's webs, for example, are natural in the sense of not being made by humans, but not natural in that they do not possess the ability to move or change themselves, at least not as nests or webs.[14] With the exception of these "natural artefacts," as we might call them, naturally occurring objects, in Aristotle's view, also possess the ability to move or change themselves. Indeed, it is because naturally occurring objects typically have this capacity for self-caused motion or change that Aristotle thinks the distinction between naturally occurring and humanly manufactured perceptible objects is not trivial. Some of his earlier contemporaries argued that this distinction merely pointed to a difference in the efficient causes that produce natural and artificial objects and was unimportant with respect to their intrinsic nature.[15] Aristotle, however, argues that the capacity of most naturally occurring perceptible objects to move or change themselves is important with respect to their intrinsic nature, and thus marks an important difference between them and artefacts.

In addition to these two ways of distinguishing natural things from non-natural ones, Aristotle sets out four more ways in which to think about what is by nature, focusing now not on nature understood as a set of objects, but on the nature of something, primarily the nature of perceptible objects.[16] In this sense, the term *nature* does not refer to a set of objects, but to a part or aspect of an object.[17] The four ways in which we can understand the nature of something, Aristotle argues, correspond to the four explanatory causes. These four causes are not restricted to changeable objects, and so when the term *nature* is understood simply as one of the explanatory causes, it can be used to refer to part of a non-changeable thing.[18] In the *Physics*, however, Aristotle is interested primarily in the nature of changeable objects, in particular, the nature of naturally occurring objects; that is, he is interested in what it is about naturally occurring objects that makes them natural in the additional sense of being able to move or change themselves.

His answer to this question focuses on efficient causes. Naturally occurring perceptible objects typically possess causal capacities by virtue of which they can initiate certain changes in themselves, and these capacities are non-accidentally located in them. The spatial location of these causal powers is crucial to Aristotle's account of natural motion; natural changes are instances of self-motion or self-change, because they are self-caused changes, that is, the efficient cause is located in the thing undergoing the change. In addition, Aristotle argues, it is characteristic of these natural causal capacities that they cannot be separated from the object to which they belong, at least not without destroying that object. By contrast, some causal capacities located in the thing undergoing the change are separable from that thing.[19] Physicians, for example, may use their medical skill to heal themselves, but physicians and their patients need not be the same person, and indeed typically are not. By contrast, the causal capacities that produce natural changes are non-accidentally located in the thing that they are moving or changing; indeed, natural substances are typically defined by these internal causal powers and are destroyed when they lose them. The nature of naturally occurring substances, then, lies primarily in their causal powers, ones that are inseparable from them and enable them to move or change themselves in certain ways.

The question, then, arises of what it is about natural substances that is responsible for these internal, inseparable causal powers. In one view, all of these causal powers belong to natural substances because of their raw materials.[20] Indeed, in this view, perceptible objects of any kind are what they are solely because of the physical raw materials from which they are made; the nature of naturally occurring perceptible objects is found in their material cause. Aristotle argues against this position by appealing to non-natural objects, namely physical artefacts: since we do not think that physical artefacts such as beds and houses are what they are solely because of their raw materials, we should not think about even more complex things such as plants and animals in this way.[21] Both naturally occurring perceptible objects and physical artefacts have a formal cause, in addition to their material cause, and this formal cause is a necessary part of what they are. The difference is that the formal cause of naturally occurring substances enables them to move and change themselves in ways over and above what they can do by means of their raw materials; the formal cause of artefacts does not.

Thus, Aristotle's claim that natural substances have two natures is not an attempt to find room for their material cause; on the contrary, he is arguing against those who would dispense with their formal cause and reduce natural substances to just their raw materials.[22] Natural substances have two natures, because their material nature is insufficient to give a complete explanation of their behaviour; a formal cause is also required.[23] Inasmuch as their final cause – the good for the sake of which they move or change themselves – typically involves the proper exercise of the causal capacities connected to their formal cause, in preserving their formal cause he is also preserving their final cause.[24] Restoring formal and final causes to their proper place in explaining the behaviour of perceptible objects, however, does not strip material causes of their own causal powers. All four explanatory causes, including the material cause, are required to give a complete account of the behaviour of natural substances.[25] Natural substances have many natures, because their nature cannot be explained just by their material cause.

7.3 The Difference between Natural Substances and Physical Artefacts

The traditional view of Aristotle's material cause, discussed in chapter 4, rejects the claim that natural substances have multiple natures, for, in this view, natural substances owe all of their necessary properties to their formal cause; their material causes either have no nature of their own or are just accidental to the substance made from them. As a result, unlike the material causes of physical artefacts, the material causes of natural substances have no independent role in determining the composition and behaviour of the objects made from them. In particular, the material causes of natural substances are responsible for none of the permanent physical attributes of these substances; all such attributes, including all of the unchanging ways in which natural substances interact with physical objects, are grounded in the formal cause of these substances. It follows that natural substances and physical artefacts are made out of their material causes in radically different ways.[26]

Aristotle, however, shows no such qualms when comparing natural substances to artefacts. On the contrary, with respect to the relation between their formal and material causes, Aristotle emphasizes the similarity between them.[27] In Book I of his *Physics*, he says that the underlying principle in natural substances must be understood by analogy; all of the examples he then gives to illustrate this underlying principle are the raw materials of physical human artefacts.[28] The same is true in Book IX of his *Metaphysics* when he explains the second, metaphysically more important meanings of "potentiality" and "actuality," the ones meant to apply to substances and their material causes. Again Aristotle says that the argument must be made by analogy, and again the examples given for comparison are physical artefacts and their raw materials.[29] It may be that these are only analogies, and that artefacts and their raw materials are not really instances of substances and their underlying principles.[30] Even if this is the case, the point of these analogies remains that natural substances and physical artefacts are made out of their underlying material causes in a similar manner. Aristotle's position cannot be that the underlying material principle in natural substances is best understood by thinking about the raw materials of artefacts, but also that this underlying principle has a completely different role in natural substances as compared to artefacts.

Furthermore, when Aristotle considers what kinds of explanation are appropriate to natural science, he again compares natural substances to human artefacts,[31] for the similarity between them justifies the use of formal and final causes in natural science, over and above the material cause.[32] Artefacts typically have a particular set of functions to perform, and the instrumental organization they require to perform these functions is not ordinarily found in their raw materials.[33] The same is true of natural substances.[34] Like the productive sciences – the sciences that deal with the production of artefacts – the natural sciences need formal and final causes to give an adequate account of the composition and behaviour of naturally occurring perceptible objects. This similarity holds, Aristotle argues, even though there is a conscious artisan at work in the construction of artefacts, but not in the generation of natural substances.[35] The upshot is that the difference between natural substances and physical artefacts is not based upon a fundamentally different role for their material causes. Both natural substances and physical artefacts require a certain kind of material cause and depend upon its causal powers. Like human artefacts, some of the permanent attributes of natural substances must be explained by their material cause, independent of their formal cause. The difference between natural substances and physical artefacts is not that natural substances have only one nature, due to their formal cause, whereas physical artefacts have two, due to their formal and material causes.[36]

Instead, as we saw above, Aristotle argues that natural substances are different from physical artefacts because they have the capacity to move or change themselves, whereas artefacts do not. More precisely, they have the ability to change themselves with respect to location, qualities, or size; thus, they are capable of self-caused locomotion, alteration, or growth and diminution.[37] With respect to locomotion, Aristotle further distinguishes between self-motion in animals and self-motion in all other natural substances: animals alone are capable of conscious self-motion.[38] By virtue of their capacity for perception and desire, animals can determine the direction of their motion, within the limits of what is physically possible for them, whereas inanimate natural substances cannot. Some natural substances, then, are capable of more complex forms of self-motion.

All instances of locomotion, alteration, and change of size are natural insofar as their efficient cause is non-accidentally located inside the thing undergoing the change. Generation, by contrast, is never self-caused.[39] Of these four types of change, generation is the only one that cannot be natural in the sense of being caused by an internal causal power. Like the production of artefacts, the generation of natural substances takes place in the raw materials and is always caused by an external agent.[40] Generation, however, is natural to the extent that the generating activity of the external agent arises from an activity internal to it; procreation is natural to the parents, not the offspring. As in the production of artefacts, the role of the material cause in generation is passive inasmuch as an external cause works upon it, imparting a formal cause.[41] This passivity in the material cause, however, is qualified. As we saw above, the relation of hypothetical necessity between a composite object and its raw materials requires the raw materials to have certain causal powers of their own. Thus, the passivity of the raw materials is restricted in that it holds only in relation to the external cause of generation. The material cause is still responsible in its own right for some of the natural, self-caused motions and changes of natural substances.

In distinguishing between natural substances and artefacts with respect to the capacity for self-motion, Aristotle is well aware that physical artefacts also move and change themselves: marble statues fall to the ground, wooden beds rot, and robots (or automata, as they were called in antiquity) make human-like motions. Aristotle argues, however, that these self-caused changes are initiated by what is natural about these artefacts – typically by the naturally occurring raw materials from which they are made – and not by whatever human artifice has added to them; the efficient cause of these self-caused changes is ultimately a natural capacity.[42] Thus, self-moving artefacts do not contradict Aristotle's claim that only natural substances can move or change themselves, for there is something natural about every artefact, and it is only by virtue of the natural part of them that artefacts have the capacity to move or change themselves, not what human artifice has added.

In Aristotle's account, then, the principal mark of natural substances is that they have an internal causal capacity that enables them to move or change themselves.[43] In this sense, the term *nature* is similar to the term *skill* (τέχνη) in that it signifies a certain kind of causal capacity and not a thing. In effect, the term *nature* refers most properly to a certain kind of efficient cause.[44] Thus, the principal difference between natural substances and artefacts, in Aristotle's account, lies in their efficient causes, not their material causes, for the distinguishing characteristic of natural substances has to do with the location and inseparability of the efficient causes moving them.[45] Moreover, unlike the production of artefacts, the efficient cause of natural generation is typically another substance of the same kind, something that already has the formal cause to be transmitted to the generated substance.[46] Finally, natural generation takes place without the accompanying activity of conscious deliberation in the agent, and sometimes without any conscious intention or desire whatsoever.[47]

The above differences all have to do with efficient causality. There is also a related difference in their formal causes. The formal causes of natural substances endow them with natures, in the sense of powers to move or change themselves; the formal causes of artefacts do not.[48] Physical artefacts may possess a second set of causal capacities by virtue of their formal cause, in addition to the ones they have by virtue of their material cause. These additional capacities, however, are passive ones, that is, capacities to be moved or used in certain ways: houses to shelter, saws to cut, dishes to contain food. The formal causes of natural substances, by contrast, are active insofar as they enable natural substances to move themselves in various ways, over and above the capacities for self-motion already found in their material causes. The formal causes of artefacts confer no such active causal powers. Thus, even the difference in the formal causes of natural substances and artefacts is grounded in the difference in the kind of causal powers that they have. In sum, the difference between natural substances and artefacts is grounded in their efficient causes, not their material causes. With respect to the role of the material cause as the substratum of change and the bearer of causal capacities in its own right, Aristotle emphasizes the similarity between natural substances and physical human artefacts.

7.4 The Unity of Natural Substances

If natural substances have a dual nature, the question arises of how to understand their unity, for Aristotle describes them as possessing a greater unity than that possessed by physical artefacts such as houses and statues.[49] This greater unity is grounded in the hypothetical necessity that holds between a natural substance and its material cause. As we saw in section 6.1, Aristotle argues that the material causes of both natural substances and human artefacts must be of a certain sort if their formal cause is to be actualized.[50] In the case of natural substances, however, the requirements placed on their material cause are greater: the demands for a certain kind of material cause go "all the way down" the hierarchy of material causes in natural substances to the very elements themselves, and the specifications for what their raw materials must be and how the latter must be combined in order to permit the actualization of their formal cause are quite demanding. At every level of their composition, natural substances have very specific requirements for their raw materials. In the case of artefacts, by contrast, the "fit" between formal and material causes is not so tight; objects such as chairs and statues can be made out of several different kinds of raw materials, and these materials often persist relatively unchanged during the processes in which artefacts are produced or destroyed. The range of raw materials out of which natural substances can be generated is typically much narrower: circles can be actualized in many different kinds of physical materials, whereas human beings can come to be in only one kind of material cause.[51]

Because the demands on the raw materials of natural substances go deeper, the changes natural substances undergo in generation and destruction also go deeper. Natural substances, for example, must revert to the basic elements before they can be used as raw materials in the generation of another natural substance.[52] As a result, it is typically more difficult to re-identify the pre-existing raw materials in the natural substance generated from them.[53] This difference, however, points to a difference in their formal cause, not their material cause. Unlike the formal causes of artefacts, the formal causes of natural substances endow these substances with the capacity to move or change themselves in certain ways. In this sense, the formal causes of natural substances do more than the formal causes of artefacts: they make natural substances more powerful in terms of their ability to change themselves. Because the formal causes of natural substances do more than the formal causes of artefacts, they also require more of their material causes.

The examples Aristotle typically gives to make this point are biological organisms.[54] Unlike the material causes of artefacts, which can be glued, tied, or nailed together, the proximate material causes of all biological organisms come to be through reproduction and growth. As we saw in chapter 6, reproduction and growth are the result of the capacity for self-nourishment and proceed by the assimilation of food by a living organism. In the course of this assimilation, the nourishment is broken down into its constituent parts, ultimately the elements. From these elements are generated the compounds and organs from which biological organisms are made. Biological organisms, then, provide the best examples of unified natural substances, because they have a number of different levels of composition, and the actualization of their formal cause, their soul, imposes demands on their material causes that go all the way down to the elements from which they are ultimately made.

This kind of unity, however, is not restricted to living beings. As we saw in chapter 6, certain elements in certain quantities are required for the generation of inanimate natural substances such as iron and marble. Thus, at all levels of their composition, natural substances must have the right kind of material cause, with the precise set of properties required to permit the actualization of their formal cause.[55] As compared to artefacts, the formal causes of natural substances impose greater demands on their material causes. In this sense, natural substances are more unified than artefacts.

The greater demands on their material causes, however, do not change the relation of hypothetical necessity between the formal and material causes of natural substances. Just because the material cause is necessary to the natural substance made from it, it does not follow that the material cause loses its own nature to the formal cause of that substance. The lesson to be drawn from the greater demands on the material causes of natural substances is not that they contribute less to the natural substances made from them, but that they contribute more. Thus, the demands on the material causes of natural substances do not entail that the latter have a single nature. Natural substances do not cease to be composite beings just because their formal cause presupposes a certain kind of material cause, for "presuppose" here does not mean "include," "explain," or "be causally responsible for." On the contrary, in order to satisfy the requirements of their formal cause, the material causes of natural substances must have certain capacities of their own, independent of their correlative formal cause. It is true that whatever acts as a material cause does so only in relation to a formal cause.[56]

Moreover, the material cause is passive inasmuch as it receives a formal cause.[57] In order to be able to receive a formal cause, however, a material cause must have certain causal capacities in its own right. Natural substances do not cease to have a dual nature just because their formal cause requires a very precise set of attributes in their material cause.

There is a further argument that often seems to lurk behind the debate about the composite nature of natural substances. According to this argument, the absence of a second, material nature is required to preserve the irreducibility of natural substances to their material cause.[58] As we saw above in section 7.2, Aristotle argues against the view apparently held by a number of pre-Socratic philosophers that perceptible objects can be understood simply in terms of the physical raw materials from which they are made. Aristotle's argument against such a reduction, however, does not require that the material causes of perceptible objects have no nature of their own or be only accidental to the objects made from them. His argument, instead, is grounded in the insufficiency of material causes to account for the patterns of motion and change that natural substances regularly manifest.[59] Thus, the irreducibility of natural substances to their material causes does not require the accidental status of their material causes. Aristotle thinks that natural substances are dependent on their material causes, but being dependent on a material cause does not entail being reducible to that cause. Natural substances are equally dependent on their formal causes, but also cannot be reduced to the latter; otherwise, they would cease to be composite objects. In general, dependence does not entail reduction. Natural substances are dependent on both their formal and their material cause, but are reducible to neither. Hence, they have a dual nature, one grounded in their formal cause and another in their material cause.

Chapter Eight

Matter and the Soul

In this chapter, we consider an argument against an Aristotelian science of matter and motion that is based on what Aristotle says about the souls of biological organisms. Aristotle argues that biological organisms are composites of a formal and material cause, and that the soul is their formal cause.[1] Thus, any general account of formal and material causes in perceptible objects must include what Aristotle says about the souls of biological organisms. This topic becomes that much more important if biological organisms are held to be the paradigmatic examples of perceptible objects; in that case, the relation between their formal and material cause will become the standard for understanding formal and material causes in all other perceptible objects. If the relation between the formal and material causes of other perceptible objects is different from that found in biological organisms, then things such as physical artefacts and non-living bodies will be understood as deviant or derivative instances of perceptible objects.

Such is indeed the view of many defenders of the traditional view of Aristotle's material cause. Biological organisms, they argue, are the paradigmatic examples of perceptible objects, because their formal cause is essential to their material cause,[2] for the parts from which biological organisms are proximately put together – their organs, limbs, and other body parts – are dependent upon the formal cause of the biological organism to which they belong. Aristotle argues that all of the organs and body parts of biological organisms are defined functionally, that is, by the way in which they are used to perform one or more of the distinctive biological functions of the organism to which they belong.[3] Given this functional definition, Aristotle concludes that the parts of plants and animals that can no longer perform their biological functions are biological parts only homonymously, that is, in name alone. When physically separated from the living organism to which they belong, or upon the death of that organism, these parts might initially appear to be the same. Without the capacity to perform their biological functions, however, they are, strictly speaking, no longer biological parts or organs. Thus, these functional parts do not persist through either the birth or death of the biological organism to which they belong.[4] They come to be during a process that begins with the generation of a living organism, and when that organism perishes, they perish as well. Thus, the functional parts of biological organisms exist only so long as the organism to which they belong is alive. In particular, these functional parts depend upon the formal cause of the biological organism to which they belong, because they are defined by the biological functions they perform, and the formal cause of that organism, its soul, specifies what those functions are.

There are, however, several problems with taking the relation between the soul and the functional parts of biological organisms to be paradigmatic for how Aristotle understands the relation between formal and material causes in perceptible objects. To begin with, if formal causes are essential to the material causes of perceptible objects, then, as we saw in the introduction, perceptible objects lack a material cause with a determinate nature of its own independent of the formal cause of the perceptible object made from it. In the case of biological organisms, if they lack such a material cause, all of their permanent features, including all of the permanent physical characteristics that they share with other perceptible objects, must be explained by their soul. As a result, the soul will be used to explain physical characteristics that have no necessary connection to life inasmuch as they also occur in many inanimate bodies. In addition, in this view there can be no single account of the physical characteristics common to all perceptible objects: in living perceptible objects, these physical characteristics will be explained by their formal cause, their soul, and in non-living perceptible objects, by their material cause or by a formal cause that is not a soul.[5] In this view, then, Aristotle cannot have a unified account of the physical characteristics found in all perceptible objects, whether animate or inanimate.

As we saw in the previous chapters, however, Aristotle does have a unified account of the physical characteristics common to all perceptible objects. Furthermore, the inability of functional parts to persist through generation and destruction raises a problem with respect to their role as material causes, for persisting through generation and destruction is a principal function of a material cause.[6] In order to act as the substratum persisting through generation or destruction, a material cause must be able to exist apart from the object made from it. Thus, the formal cause of a generated object cannot be essential to its material cause. Given Aristotle's functional definition of biological parts, however, these parts cannot act as the substratum persisting through the generation and destruction of biological organisms. Something else, other than the functional parts of plants and animals, will have to be found to persist through their birth and death.

As we saw in chapter 4, the traditional view of Aristotle's material cause solves this problem by arguing that biological organisms have a material cause that persists through their generation and destruction, but without an independent nature of its own.[7] In contradistinction to the raw materials of physical artefacts – say, the wood that is made into a wooden box – true perceptible substances in this view, including animate ones, require a material cause essentially informed by their formal cause. Thus, the pre-existing raw materials cannot survive in the biological organisms generated from them, for these pre-existing raw materials are not informed by the formal cause of that organism. As a result, the pre-existing raw materials out of which biological organisms are made and the material causes of which these organisms consist while they are alive are two different things. Once again, we are back to the problem of a persisting substratum that does not persist as anything.

The remainder of this chapter argues that the traditional view of Aristotle's material cause is not supported by his account of the role of material causes in biological organisms. While the functional parts of biological organisms are indeed dependent upon the formal cause of the organism to which they belong, biological organisms are still composite entities, with a discrete, independent material cause.

8.1 The Soul as the Actuality of the Body

At the beginning of Book II of his *De anima*, Aristotle says that the soul is a substance.[8] He then lists three different ways of being a substance: as a formal cause, a material cause, or a composite of a formal and material cause. Arguing by elimination, he concludes that the soul is a substance in the first way, as a formal cause. More precisely, the soul is the formal cause of a natural body with the potential for life.[9] A natural body with the potential for life, however, is one that already has a soul.[10] Thus, if the soul is the formal cause of a natural body with the potential for life, the natural body of which the soul is a formal cause must already have a soul. If, in turn, "natural body with the potential for life" is understood to refer to the proximate material cause of a living being, the soul will turn out to be essential not only to living beings, but to their proximate material cause as well. It seems, then, that perceptible living beings do not have an independent material cause, because their proximate material cause is itself defined by their formal cause, their soul.

Still, even if the soul of a biological organism is essential to that organism's proximate material cause, nothing in Aristotle's definition of the soul makes the other, lower-level material causes of biological organisms dependent for their own nature upon the soul of the organism in which they are found. Indeed, it is not even clear that Aristotle's definition of the soul makes much of a point about the material causes of biological organisms at all, for the expression "natural body with the potential for life," which he uses in his definition of the soul, refers more plausibly to a living being as a whole, not just to its material cause. When Aristotle comes to define the soul, he does so by looking at perceptible beings that are already alive. More precisely, he proceeds by way of division: he begins with bodies and then separates off natural bodies.[11] He then subdivides the class of natural bodies into those that have life and those that do not, where the criterion for having life is the capacity for self-nourishment, growth, and decay. Thus, Aristotle's division of natural bodies into living and non-living bodies does not presuppose any particular account of the soul; it presupposes simply that there are certain functions that living bodies can perform and non-living ones cannot. He also presupposes that all and only living beings have a soul. Thus, any natural body that can perform certain biological functions is already alive and already has a soul. The point of Aristotle's definition of the soul, then, is not to mark the difference between living and non-living bodies; he has already done that. What he wants to make clear, instead, is just what it is that a living body has when it has a soul. Aristotle uses the distinction between living and non-living natural bodies to get at the nature of the soul, not the converse.

Thus, when Aristotle enquires into the nature of the soul by looking at living natural bodies, he is looking at things in which a soul is already present. To get at the soul, he has to disentangle it from the living body in which it is already found. Consequently, when he defines the soul as the formal cause of a natural body with the potential for life, the natural body in question is not just the material cause of a living being. Natural bodies with the potential for life are already living beings and already have a soul.[12] In defining the soul this way, Aristotle is first identifying living beings and then specifying what part of them is their soul. He is determining the relation between the soul and a living being, not the relation between the soul and the material cause of a living being.[13]

Thus, neither implicitly nor explicitly does Aristotle's definition of the soul make the material causes of living beings dependent on their soul; if anything, the opposite is the case, for the process of division that leads to Aristotle's definition of the soul presupposes an independent material cause. The soul is the formal cause of a natural body of a certain kind, one with the capacity for life, but the soul does not make that natural body to be a natural body in the first place. Aristotle looks at natural bodies first, then living natural bodies. The presence of the soul makes the difference between living and non-living bodies, but it does that by conferring on natural bodies an additional set of powers over and above the ones that already belong to them as natural bodies. The corporeal nature of living beings is presupposed, not established by the soul.

This account is confirmed by Aristotle's treatment of biological organs in his *De anima*. Here he typically first identifies an organ and then asks what part of that organ is its formal cause. In the case of the eye, for example, its formal cause is sight, the capacity to see.[14] This formal cause is not added to an eye; to be an eye, a body must already be capable of seeing. To get to the material cause of the eye, sight has to be taken away from the eye, but the material cause of an eye is no longer an eye. Thus, when Aristotle says sight is the formal cause of the eye, he is not talking about the relation between a formal cause and a material cause; he is talking about the relation between a formal cause, sight, and the thing of which that formal cause is already a part, the eye. The same holds for his treatment of the other organs. Far from emphasizing their role as material causes, Aristotle emphasizes the similarity between living beings and their functional parts. For the most part, he treats functional parts not as material causes, but as mini-organisms, in which a biological formal cause is connected to an inanimate material cause. Thus, Aristotle's definition of the soul does not call into question the independent status of the material causes of living beings. If anything, it is the status of their functional parts as material causes that is uncertain., for he says that the soul must be taken away in order to get to the material cause of living beings.[15] Taking away the soul, however, leads to physical matter, not functional biological parts.

As we saw in section 3.1, Aristotle allows something that acts as a material cause for something else to have a material cause of its own. The resulting layering of material causes is especially clear in the case of biological organisms. Here Aristotle distinguishes three general types of material cause: (1) their organs and larger body parts; for example, hands, eyes, lungs, and teeth in animals, and roots, leaves, stems, and flowers in plants; (2) homogeneous mixtures such as flesh and bone, from which their organs and larger body parts are made; and (3) the material elements, from which all homogeneous mixtures are made.[16] In this list, the organs and larger body parts are the proximate material cause of biological organisms; homogeneous mixtures such as flesh and bone are non-proximate, more remote material causes; and the material elements are their ultimate, separately existing material cause. There are, however, important differences between these three layers. The first consists of the functional parts of biological organisms; as we saw, they are defined by their ability to perform specific biological functions and, in this sense, are dependent upon the formal cause of the biological organism to which they belong. The second layer consists of homogeneous mixtures. Part of them is determined by the biological organism to which they belong, but part of them is independent of that organism and its formal cause. Human bones, for example, have certain characteristics because they belong to humans, but part of them is independent of not just a human organism, but of any biological organism; as we saw in chapter 6, bones harden in the same way as earthenware vessels. The third layer consists of the material elements, and they are what they are and behave as they do entirely independent of the perceptible objects made from them. Thus, in addition to their functional parts, biological organisms have two other layers of material causes that are – either in whole or part – independent of the formal cause of the organism made from them.

This progression from dependent to independent material causes is not unique to biological organisms. The functional parts of physical artefacts are also defined by the formal cause of the object to which they belong.[17] They too cease to be parts of the artefact to which they belong when they lose their functional capacity; thus, they too cannot persist through the construction or destruction of that artefact. All perceptible objects, however, also have material causes that are independent of the formal causes of the objects made from them. Functional parts do not take over responsibility for the causal capacities of these lower-level material causes or render them superfluous. Indeed, precisely because functional parts are dependent upon the formal cause of the object made from them, they cannot supply the causal capacities and attributes that composite objects require from their material causes. Far from being exemplary material causes, the functional parts of perceptible objects are only imperfect examples of material causes. In fact, if, as Aristotle indicates, the principal role of the material cause is to persist through generation and destruction, functional parts are not material causes at all. They are parts of the composite object, not of its material cause.

In sum, Aristotle has strict requirements for functional parts: parts that cannot perform their functions are functional parts in name alone. The independence of the material causes of perceptible objects is not undermined by this claim. When Aristotle invokes his homonymy principle that the non-functioning organs of a biological organism are organs in name only, he is refusing to attribute a formal cause to these organs, not a material cause. It is their formal cause that these parts no longer possess; their material cause remains. In fact, their material cause is all that is left to these lifeless organs.[18] Aristotle's homonymy principle does not claim that living organisms have no independent material cause or that the physical matter of biological organisms contributes nothing permanent to them. His claim that the functional parts of biological organisms do not persist through their birth and death is entirely consistent with the claim that their lower-level material causes do.[19] Moreover, through all of these biological processes, living beings and their functional parts remain subject to, and dependent upon, the physical necessity that governs the elements and everything made from them. Once again, then, natural substances, including biological organisms, are radically composite objects, with two natures, one grounded in their formal cause and another in their material cause.

8.2 The Soul and Efficient Causality

The same result is reached if we consider the relation between the soul and efficient causality. Aristotle says that the soul is an efficient cause, as well as a formal and final cause.[20] As we saw in section 2.1, Aristotle speaks of both agents and causal powers as efficient causes, but causal powers do not operate independently of agents. Causal powers need to be included in causal explanations, because it is not enough to specify just the causal agent; the parts and capacities by means of which the causal agent produces a change must be included in a complete explanation. The soul is relevant to causal explanation precisely in this role of causal power, for the soul is a formal cause that endows the living being to which it belongs with a set of powers.[21]

These causal powers, however, must be found in a body of a certain kind. Aristotle repeatedly states that biological organisms are composite beings, consisting of a soul and a body, and it is this composite being that performs their distinctive functions, not the soul by itself.[22] With the possible exception of the agent intellect in human beings, the causal capacities of biological organisms belong to composite beings, consisting of a formal and material cause, not to their soul or formal cause alone. The causal capacities of biological organisms, in turn, are connected to their organs, limbs, and functional parts, for, as we saw in chapter 6, Aristotle thinks that biological capacities can be found only in perceptible objects that are structured in the requisite manner. Biological functions, beginning with reproduction and self-nourishment, can be performed only by objects with the right functional parts. These functional parts, in turn, require certain material causes of their own, for the biological functions of living organisms and their parts also presuppose the non-biological causal capacities of the material causes from which they are made. Thus, if these physical material causes are taken away, biological organisms lose their biological powers as well.

As a result, the functional parts of biological organisms are not sufficient to make biological organisms into composites of a formal and material cause. There has to be something to these functional parts other than what their soul makes them to be. Aristotle's homonymy principle implies that functional parts of biological organisms do not have a formal cause of their own, but are defined by the soul of the organism to which they belong. Thus, without their lower-level material causes, the functional parts of a biological organism turn out to be just parts of that organism's soul. If biological organisms are to be composites of a formal and material cause, their functional parts must also include material causes that are not defined by their soul. There has to be more to their body than just their soul.

In fact, as we saw in chapter 6, the functional parts of biological organisms depend upon their lower-level physical material causes in order to perform their biological functions. Animate bodies depend upon inanimate ones. This dependence is not refuted by the fact – of which Aristotle was well aware – that biological organisms repeatedly exchange their physical matter during the course of their life, for this exchange shows that biological organisms still need their lower-level material causes. While the physical matter of biological organisms typically does not remain numerically the same during their lifetime, at every moment of their life biological organisms must have some physical matter in order to perform their biological functions. In addition, as we saw in chapter 6, the physical matter from which biological organisms are made must always remain sufficient in kind and quantity to instantiate the formal causes of these organisms; not just any kind or amount of matter will do. Finally, in every exchange of physical matter, some of it must persist; a biological organism cannot exchange all of its physical matter at once. Thus, although the material elements and the homogeneous mixtures made from them are constantly being replaced in living biological organisms, they are constantly being replaced with other elements and homogeneous mixtures, and the organism doing the replacing must itself contain this kind of physical matter. The soul is not always found in the same bit of physical matter, but it must always be found in some physical matter. What persists through the life of a biological organism is the composite object consisting of a soul and physical matter, and that remains true even if its physical matter is constantly being exchanged, bit by bit.

To conclude, the role of their functional parts as proximate material causes does not make the lower-level material causes of biological organisms superfluous or accidental to them. The functional parts of biological organisms are dependent upon the soul of these organisms, but this dependence does not make biological organisms independent of their physical matter. On the contrary, although biological organisms are immediately made out of their functional parts, many of their necessary attributes are due to the causal powers of the physical matter from which they are ultimately made. Moreover, biological organisms depend upon this physical matter in order to perform their own biological functions. Part of the regular behaviour of biological organisms, then, is explained by looking "downward" to their physical matter. Not only is there is room for physics in Aristotle's biology, but the latter presupposes the former. This claim, applied to all functionally organized perceptible objects, is the subject of the next chapter.

Chapter Nine

The Role of Teleological Explanation

In chapter 8 we saw that Aristotle's account of the soul of biological organisms is compatible with a science of matter and motion. This chapter argues that the same is true of his use of teleological explanation. Aristotle's view that many changes in the physical world happen the way they do because it is good, or even best, that they happen this way is perhaps the most controversial part of his account of the material world. Once the limits of Aristotle's use of teleological explanation are understood, however, there is no inconsistency between Aristotle's use of final causes in his biology and elsewhere, and his account of the non-teleological physical necessity operating in all perceptible objects. Indeed, as we shall see, the former presupposes the latter.

9.1 The Limits of Teleological Explanation

The first limit on the use of final causes has to do with their role in the explanation of change. Aristotle defines a final cause as the good for the sake of which a change takes place.[1] Thus, like efficient causes, Aristotle explicitly connects final causes to the occurrence of change. Hence, where there is no change, as in the objects of mathematics, there is no place for final causes.[2] Aristotle also contrasts final causes with efficient causes; final causes are not efficient causes and, by themselves, produce no change.[3] As we saw in section 2.1, Aristotle insists that efficient causality can be exercised only by physical causal agents. In particular, all efficient causality in perceptible objects requires reciprocal contact; consequently, the proximate mover is always moved by the thing it sets in motion.[4] Precisely because final causes are not efficient causes, they can move something else and yet remain unmoved themselves.[5]

As objects of desire, for example, final causes move things that desire them, but this is not efficient causality; the role of the efficient cause here belongs to the agent desiring the final cause.[6] To the extent, then, that final causes depend on change, and change depends on efficient causes, final causes presuppose efficient causes. The first limit on teleological explanation, then, is the limit on what final causes can do, or rather, cannot do, namely produce a change; for that, an efficient cause is required. Stated in more contemporary language, final causes are effects, not causes.[7]

To be sure, as an effect produced by a causal power, final causes are closely connected to efficient causes. In particular, final causes tell us something about the nature of the causal powers that produce them, for only certain sorts of causal powers can produce certain sorts of effects. As we saw in section 2.1, if a causal agent regularly produces a beneficial result – or, indeed, regularly produces a result of any kind – it needs to possess the appropriate internal structure and causal capacities. In particular, Aristotle argues that regularly occurring beneficial results cannot be the result of chance or accident; they require causal agents of a certain kind.[8] Thus, even though final causes are not efficient causes, they can tell us much about the causal agents that produce them.

Those causal agents need not be conscious agents. Although changes arising from the conscious pursuit of an end are perhaps the most accessible examples of changes that have a final cause, they constitute for Aristotle only a small subset of teleological changes; the latter include everything from biological reproduction to the natural motions of physical bodies.[9] The natural motions of the sublunary elements, for example, are good inasmuch as they are eternally ordered, but they involve no conscious desire. There is also no necessary connection between final causality and contingency, that is, the possibility that the result could have been otherwise. The motions of the stars are good, in Aristotle's account, because of the perfection of their circular pattern, but they are also eternal and, given the causal forces involved, cannot be otherwise. Thus, thinking about final causes primarily as the contingent objects of choice is misleading.

The second limit on the use of teleological explanation arises from Aristotle's claim that final causes are the good for the sake of which a change occurs; more precisely, a change has a final cause only if it has a beneficial result and the stages of that change are ordered by what is required to reach that result.[10] Thus, a final cause is more than just the termination of a change; otherwise, Aristotle jokes, death would be the final cause of life.[11] In addition, the result of the change has to be beneficial either for the thing undergoing the change or for something else affected by it.[12] Because such beneficial results are widespread in nature, Aristotle thinks that final causes must be included in any complete explanation of the physical universe.[13] In particular, he applies teleological explanation to all natural substances inasmuch as he claims that all natural substances engage in activities the complete explanation of which requires final causes.[14] Nevertheless, as we saw in the earlier chapters, it is not the case, in Aristotle's account, that everything that natural substances do can be explained teleologically. It may be that, given the structure of the physical universe, every natural change ultimately benefits something or other, but it is not the case that every natural change has its own final cause.[15] Natural substances do many things, and not all of them are for the sake of an end. In particular, where a change does not result in anything that is good or beneficial, there is no need for a final cause.

Stated in terms of Aristotle's definition of change, it may be that every change is the actualization of a potentiality, but sometimes what is being actualized is not to the good of the object undergoing the change.[16] In some cases, this is because what is being actualized is indifferent to the good of the changing object. This indifference holds particularly for changes in which nothing is generated or destroyed, that is, instances of alteration, change of size, and locomotion. Moreover, sometimes the actuality coming to be is harmful to the thing undergoing the change, such as decay or destruction.[17] Death actualizes a biological potentiality, but it is not the good for the sake of which biological organisms exist or are generated. Not every change, then, actualizes something good or beneficial. Thus, it is not part of Aristotle's definition of change that every change has a final cause.

Eclipses are one example given by Aristotle of changes that occur regularly but do not have a final cause.[18] To be sure, eclipses occur as a result of the motions of the planets and stars, and these motions have final causes. Many other changes in perceptible objects, however, lack a final cause.[19] Perhaps most important, no final causes are required to explain the lowest level of simple physical necessity at work in the material elements. As we saw in chapter 5, this kind of physical necessity is non-teleological because the basic changes in the elements do not take place for the sake of a good. The defining capacities of the sublunary elements consist in the way in which they interact with other physical bodies. The exercise of these capacities produces no benefit for these elements; there is nothing about being warmer or colder, more or less rigid, that is intrinsically good for the elements. Otherwise, it would be bad for the sublunary elements to be combined into homogeneous mixtures, and Aristotle gives no indication that this is the case. The processes of heating and cooling can lead to the generation and destruction of the elements, but many interactions between the elements and physical bodies do not; when a warmer body is cooled by a cooler body through physical contact, no generation need occur in either body, and neither the cooling of the one body nor the warming of the other need be beneficial to either of them. A warm stone is cooled through contact with a cold one, but neither is benefited or harmed by these changes as such.

Moreover, the operation of the basic causal powers of the sublunary elements is different from events such as eclipses inasmuch as eclipses are derivative occurrences, arising out of the natural motions of the planets and stars. In the case of the sublunary elements, however, their tactile powers are what define them, and no end or intrinsic good is reached through the exercise of these powers by themselves. The same holds for the collisions between bodies that produce sound or the transmission of light by transparent bodies, causal processes that involve both celestial and terrestrial elements. There is, then, nothing beneficial, nothing good or bad, about the fundamental operations of the elements.

It follows that teleological explanations do not go "all the way down" in explaining the behaviour of natural substances at all levels. In fact, as we saw in chapter 6, the non-teleological necessity at work in the elements governs all other perceptible objects as well, because the latter are all ultimately made out of the elements.[20] Thus, the simple, non-teleological physical necessity that governs the five material elements goes "all the way up" in determining, at least in part, the behaviour of all of the more complex physical objects made from them.[21] In effect, the non-teleological physical necessity in the elements extends to everything made out of them.[22] As a result, all goal-directed behaviour in natural substances is limited by the non-teleological necessity at work in the material elements. The second limit on the use of teleological explanation, then, is that not every motion or change is for the sake of a good. Where there is no beneficial result, there is no final cause.

The third limit on teleological explanation has to do with which efficient causes are directed towards the production of a final cause. Not only do some changes not have a final cause, but even when there is a beneficial result, this outcome does not always mean that the efficient cause that produced it was directed towards producing that effect. Efficient causes, according to Aristotle's, are understood primarily in terms of the effects that they produce. Strictly speaking, however, this connection holds only when the efficient cause regularly and directly produces that result. As we saw in chapter 2, Aristotle argues that when a certain effect – including a beneficial one – is regularly produced by an efficient cause, that result cannot occur accidentally or by chance. Thus, when a beneficial result is regularly produced by an efficient cause, Aristotle holds that the efficient cause in question acts for the sake of that result, again without any requirement for a conscious desire. When a beneficial result occurs by chance, however, the efficient cause that produced this beneficial outcome does not act for the sake of that result; the beneficial result comes about through the concatenation of several efficient causes, none of which was directed to that end.[23] In the case of chance events, the connection between cause and effect is accidental, because it is grounded in the accidental properties of the things involved; in these cases, the nature of the effect tells us little about the nature of the cause. Chance events have a beneficial result, but not an efficient cause acting for the sake of that result.

Similarly, the efficient cause must be adequate to produce the beneficial effect by means of its own causal powers before it can be understood to be directed to that end, for every efficient cause can be made to produce a good result in the right context. Although there is nothing teleological about the basic processes that occur in the elements, it is always possible to introduce a beneficial outcome when these elemental changes take place as part of a larger process. The operation of heat, for example, is not teleological as such, but it can be used to produce many results beneficial to human beings, everything from heating our dwellings in cold weather to cooking our food. Thus, to show that a causal power is exercised for the benefit of an object, it is not enough to show that some benefit will be produced by the exercise of that causal power; a final cause can always be added by putting something to work for the benefit of something else.

Thus, to prove that a causal power is directed towards a final cause, not only must the causal power produce a beneficial result non-accidentally, but it must also be able to produce that result without being added to another, goal-directed process. As a result, in objects composed of a formal and a material cause, for the causal capacities of the material cause to produce a result beneficial to the composite object as a whole, those material capacities need not be explained teleologically.

The hardness of the metal in a saw blade is useful for cutting wood, but the metal is not hard for the sake of cutting wood. The job of cutting wood belongs to the saw, not to the metal from which it is made. The teleology in the composite object does not exclude non-teleological causal powers in its matter.

In sum, the reach of teleological explanation, in Aristotle's account, does not extend to everything that happens in the physical universe. Final causes depend upon efficient causes, apply only to changes that have a beneficial result, and are evidence of a goal-directed efficient cause only if the cause and effect are non-accidentally connected. Many changes fail to meet these requirements. It may be that teleology is universal in nature in that every natural substance, at some level, acts for the sake of some end or other. Teleology, however, is not universal in nature in that not every causal power in natural substances is exercised for the sake of a final cause.

9.2 The Compatibility of Natural Teleology with Non-Teleological Necessity

In chapter 6 we saw that the two kinds of simple physical necessity operating in the material elements go "all the way up" in the sense that they also operate in everything made out of the elements. One of these types of physical necessity – the one governing the physical interaction of the material elements – operates in a non-teleological way. How, then, is this non-teleological physical necessity compatible with the goal-directed activities of natural substances? This question also arises for the other, teleological type of physical necessity operating in the elements – the one governing their natural motions – because this second kind of physical necessity also limits the specific goal-directed activities of natural substances: whatever their own ends may be, these ends must be pursued within the limits of what is physically possible, which is set by the elements from which all perceptible objects are made.

Aristotle's solution to this problem is grounded in his claim – discussed in chapter 7 – that natural substances have two natures, one grounded in their formal cause and another in their material cause, for, in Aristotle's account, final causes are grounded in the formal causes of changeable objects, not in their material cause.[24] Given the role of formal causes in determining the specific nature of perceptible objects, this connection between final and formal causes is not surprising, for final causes are based on what is beneficial for something, and the latter, in turn, is determined by what kind of thing that object is.[25] In the case of biological organisms, for example, their formal cause enables them to perform their distinctive biological functions; thus, generation and growth are beneficial for biological organisms because their formal cause is actualized by these changes.[26] Even after a formal cause has been brought into being, the final cause of a perceptible object is still connected to its formal cause, for generally the exercise of some faculty or other is beneficial to a perceptible object if it is part of the proper exercise of that object's distinctive capacities, and those capacities are introduced through the addition of the object's formal cause, not the capacities of its material cause alone.[27]

As a result, the limits on the formal causes of perceptible objects also limit their final causes. The formal cause of a perceptible object presupposes a certain material cause but does not explain the causal capacities of that material cause. The same restriction applies to final causes. Final causes do not explain everything about perceptible objects, because formal causes do not explain everything about them. In the previous section, we saw that not every formal cause is suitable to be a final cause; where there is no beneficial result, there can be no final cause. The dependence of final causes on formal causes imposes a further restriction on teleological explanation: even where the formal cause does specify what would be beneficial for a perceptible object, the formal and final causes will still not explain everything about that object.

Thus, the non-teleological physical necessity at work in the material elements is compatible with the goal-directed activities of the composite perceptible objects made from them, because the final causes at work in composite objects are grounded in the formal causes of the latter, and these formal causes are added to the material causes of composite objects. Formal and final causes supplement the material nature of perceptible objects; they do not replace it.[28] Final causes do not compete with material causes in perceptible objects, because these two causes explain different parts of the composition and behaviour of these objects. It is not the case that the same features of perceptible objects are explained in two different ways, one teleological and one non-teleological.[29]

Here again it is important to remember that the functional parts of perceptible objects are not, strictly speaking, material causes. As we saw in chapter 8, the functional parts of perceptible objects are defined by the function or functions of the composite object to which they belong. Thus, they are parts of that object as a whole, not of its material cause.

The organs of plants and animals, for example, are parts of living organisms, but not parts of either the formal or material cause of those organisms. The same is true of the functional parts of artefacts; they too are defined in relation to the function of the artefact to which they belong and require both the formal and material cause of that object in order to perform their own functions. Moreover, functional parts depend upon the continued functioning of the object to which they belong; unlike material causes, they cease to exist if the composite object to which they belong can no longer function. The elements and non-specific parts of homogeneous mixtures are not functional parts of the perceptible objects made from them, because they are not defined in relation to the functions of those perceptible objects. Thus, they can persist when the composite objects made from them perish.

In sum, the teleological and non-teleological changes that take place in perceptible objects are compatible because final causes and material causes explain different features of the physical world. In particular, the basic, non-teleological causal properties of the elements are indifferent to the teleological features of the composite objects made from them. As the production of artefacts makes clear, the necessary movements and changes of the elements do not exclude more complex forms of order; physical necessity in nature leaves room for contingency. Within the boundaries set by physical necessity, there is much that is neither contrary to, nor follows from, the basic attributes of the elements and the other, non-specific material causes from which changeable objects are made. Because of this indifference to higher levels of order, the material causes of perceptible objects cannot explain the instrumental organization and purposive activities of the objects made from them. Indeed, as with artefacts, it is precisely because of the indifference of the material causes of natural substances to the instrumental features of the latter that additional causes, over and above their material causes, are required to give a complete explanation of how natural substances behave. Moreover, with respect to the instrumental organization of a natural substance and the process by which its formal cause is brought to be in its material cause, the final cause comes first.[30] There is more to natural substances, however, than just their formal cause.

Understanding final causes as grounded in the formal causes of perceptible objects also explains how Aristotle can extend teleology beyond individual perceptible objects. As we saw above, the parts of a perceptible object are for the sake of the good of that object only when they are functional parts whose activities in some way benefit that object. Aristotle also applies this type of teleological explanation to the interaction of several individuals when they are part of some larger whole. He argues, for example, that rainfall, at a certain time of the year, is for the sake of the growth of crops, that is, plants suitable for human consumption.[31] He also describes animals as existing for the sake of human beings inasmuch as they too supply us with a source of food.[32] In both cases, one kind of thing benefits another in that it enables the latter to perform one of its distinctive functions. Thus, Aristotle sees teleology as explaining not only the relation between functional parts and the object to which they belong, but also the relation between several objects when together they constitute a functional whole.

Some commentators have argued that this extension of teleology to multiple objects is illegitimate.[33] When it comes to deciding whether a change can be explained teleologically, they argue that only individual objects should be considered; when a change taking place in one object is beneficial for another, spatially discrete object, it is not the case that the change in the first is for the sake of the second. In effect, teleological changes must be internal to the thing being benefited. As a result, final causes are applicable to the relations between parts and wholes only when the parts are internal to the thing being benefited, such as the organs of biological organisms.

The advantage of this internalist account is that it prevents changes in perceptible objects from being explained by irrelevant and misleading factors. The problem with the externalist account is that it seems to explain certain natural events, say, the falling of rain, in terms of other natural processes, say, the growing of plants, where the former have no necessary connection to the latter. After all, rain often falls where there are no plants, say, on rocks or rooftops, or during a season when food plants no longer grow. Some commentators have tried to make this connection more plausible by arguing that it is not rainfall as such that Aristotle is explaining teleologically, but rainfall at the time of year when it is needed by plants.[34] This account, however, still seems strained, because the explanation of rainfall should be the same whenever it occurs.

This objection to the externalist account, however, is blocked when we remember that teleology does not explain everything about the structure and behaviour of perceptible objects. In particular, with respect to the parts of functional wholes, regardless of whether those functional wholes are individual objects or a causally connected group of objects, final causes explain only the part of these objects that is connected to their specific function. Thus, final causes explain the formal structure of functional parts but not the causal capacities of their material causes. These lower-level material causes may be necessary for the proper functioning of the object made from them, but their own behaviour is not explained in terms of the function of that object. Similarly, when applied to a group of spatially discrete objects, final causes explain why the behaviour of some objects in that group is useful to others, but not why those individual objects behave that way in the first place. In individual perceptible objects, final causes operate only above the level of the causal capacities already found in their material causes. In causally connected groups of individuals, final causes explain only how the behaviour of one thing contributes to the functioning of something else; they do not explain why the first thing behaves the way it does, independent of its contribution to the larger whole. The formal and final causes of plants may explain the benefit of rainfall for them at certain times of the year, but they do not explain rainfall as such; the latter is always and everywhere explained in the same way, independent of any further benefits or harms that may come from it.

Thus, final causes need not be restricted to internal teleology; the activity of one thing can be understood to be for the sake of something else where the activity of the first regularly produces beneficial consequences for the second. In both individuals and groups of individuals, however, the final cause explains only the additional or incremental effect, that is, the beneficial consequences that the activity of one thing regularly produces for something else. Final causes do not explain the lower-level causal powers that parts or individuals possess independent of the larger whole to which they belong. Given their composite nature, all perceptible objects have at least one, and typically several such lower-level material causes that have their own causal powers independent of the composite object made from them. The limits to a teleological account of nature, then, are set by these lower levels of physical matter; no perceptible object is ever wholly for the sake of its own final cause or the benefit of something else. In sum, both the internalist and externalist accounts of teleology are restricted in what they can explain. Neither within nor between perceptible objects do final causes explain the lower-level causal capacities that they presuppose.

Understanding final causes as grounded in the formal causes of composite objects, but not their material causes, also explains how Aristotle can understand certain events as happening both of necessity and for the sake of an end. In the *Posterior Analytics*, for example, he says that, of necessity, the light from a flame passes through the pores in a surrounding screen that are large enough to permit their passage, but when this flame is in a lantern held aloft by someone walking in the dark, the passage of the light through the lantern's screen helps that person avoid stumbling.[35] In the case of rainfall, the water falls to the ground both of necessity and for the sake of its contribution to the growth of plants.[36] The same thing happens internally, within complex perceptible objects. As we saw in chapter 6, processes such as heating and cooling, liquefaction and solidification, take place within biological organisms by physical necessity, but, inasmuch as the results of these processes are useful to the biological organisms in which they take place, they also happen for the sake of an end.[37] In these and other examples, a change takes place both of necessity and for the sake of an end, but not in the same respect. At a lower, material level, something can happen of necessity, for no end, but then that change can also be useful when it takes place within the context of a larger, functional whole. Again, it is not the case that the same event or the same feature of a perceptible object is being explained in two different ways. The teleological explanation in these cases brings in other objects or other features of the same object and considers the contribution of the lower-level entities and their causal powers to a more complex object or group of objects.

Finally, if final causes explain only the role of functional parts, but not the material causes of those functional parts, the compatibility of the two types of unconditional physical necessity operating in the elements also becomes clear. As we saw in chapter 4, the necessity governing the natural motions of the elements is teleological, whereas the necessity governing their physical interactions is non-teleological. These two types of necessity are compatible because the teleological motion of physical bodies to their natural places depends upon those bodies being part of a larger whole. As Aristotle makes clear in his arguments against the existence of a void, the material elements have their natural motions only within the context of an ordered cosmos; if the elements were placed in a void, they would lose their natural motions.[38] Considered in their own right, the elements are subject only to non-teleological unconditional necessity in their physical interactions with other bodies. Considered in relation to the eternally unchanging cosmos and their natural motions within it, the elements are subject to a second kind of unconditional necessity, this time teleological.

The more general lesson to be drawn from this account of final causality is that when explaining the behaviour of perceptible objects it is important to identify not only the correct perceptible object to which a causal power belongs, but also the correct level in that object's composition. Because perceptible objects can have multiple levels of composition, with multiple levels of material and formal causes, a causal power operating at one level might not have a final cause, but might be part of some larger object that does; the purpose of the saw does not belong to the hard metal from which the saw is made. Thus, part of the difficulty in any scientific inquiry lies in determining at what level the effects under investigation are to be explained. Aristotle argues that scientific knowledge is grounded not just in true first principles, but in ones appropriate to the subject matter.[39] We now see that the behaviour of a perceptible object is also not correctly explained just by citing universal laws that apply to that object; in addition, these laws must be applied at the right level of the composition of that object. Final causes, in particular, have to be applied at the right level. The concoction, for example, taking place in the digestive tract of an animal is beneficial to the extent that the animal is nourished by it, but there is much about this process that is indifferent to the welfare of the animal, or indeed, the welfare of anything. Because perceptible objects have multiple layers of causal powers, scientific explanation requires accurate identification of not just the correct causal agent, but also the correct part of the causal agent.

In sum, the teleological behaviour of natural substances is compatible with non-teleological physical necessity because there is much about the arrangement of their raw materials that is physically contingent; matter by itself is indifferent to many different possible arrangements and outcomes. The purposive workings of perceptible objects must take place within the limits of what is allowed by simple, non-teleological necessity, but within those limits, Aristotle argues, some changes take place the way they do because they are beneficial to the things involved. This explanatory division of labour does not lead to the over-explanation of the material causes of perceptible objects, because the beneficial use of the capacities found in these material causes does not mean that these capacities themselves can be explained teleologically.

9.3 The Dependence of Natural Teleology on Non-Teleological Necessity

Although the non-teleological physical necessity at work in perceptible objects is indifferent to their teleological behaviour, the converse does not hold. As we saw in chapter 6, the teleological behaviour of perceptible objects is dependent upon non-teleological physical necessity inasmuch as the former must make use of the latter. Given the hypothetical necessity that holds between perceptible objects and their material causes, perceptible objects can perform their own, goal-directed activities only by using the causal capacities found in their raw materials. The bones of animals, for example, have to be made of rigid materials in order to support their fleshy parts; similarly, parts such as the hoof, claw, beak, and teeth have to be made of something earthy and solid in order to serve as weapons.[40] As we saw in chapter 6, if animals are to be keen-sighted, the fluid in their eyes must be transparent.[41] The transparency of the fluid in the eye, in turn, is dependent on the quantity of water in the eye.[42] Just as a saw must be made out of something hard if it is to cut wood, so too must the functional parts of natural substances be made out of suitable raw materials if they are to perform their distinctive functions.[43] In all of these instances, a perceptible object is dependent upon its material cause, not the converse.

Hence, the teleological part of Aristotle's natural science cannot replace, and was never intended to replace, the part of his natural science that deals with the basic material constituents of perceptible objects and their motions and causal powers. Apart from what the material elements do as the material causes of other, more complex perceptible objects, they constitute a natural kind by themselves, and it is necessary to understand the essential features of the material elements if one is to understand what else is required, over and above them, for more complex perceptible objects to act for the sake of beneficial ends. Far from excluding non-teleological physical necessity from nature, the teleological part of Aristotle's science of nature presupposes it.[44]

Chapter Ten

Conclusion: The Independence of the Material Cause

Aristotle's investigation of the physical world typically begins with his analysis of spatially discrete, functionally organized perceptible objects, everything from plants and animals to planets, stars, houses, statues, and drinking cups. All of these objects he sees as composite beings, consisting of a formal and a material cause. This book has considered the explanatory division of labour between these two causes. We have seen that Aristotle sees both material and formal causes as responsible for non-accidental aspects of the behaviour of perceptible objects. In particular, the material cause of every perceptible object must have properties and causal capacities of its own, independent of that object's formal cause, in order to be able to perform its various functions in the generation, composition, and operation of perceptible objects. The fundamental mistake, then, in the traditional account of Aristotle's material cause is to make material causes dependent for their own nature upon the formal causes of the objects made from them. The traditional view arrives at this conclusion on the basis of its claim that the formal cause of a perceptible object is essential to that object's material cause. This view reduces the material cause to a set of accidental properties inhering in an object whose nature is due entirely to its formal cause.

10.1 The Material Cause and the Substratum of Generation

In fact, given Aristotle's analysis of the requirements for motion, change, and generation, the formal cause of a perceptible object must be accidental to that object's material cause, for, if the formal cause were essential to the material cause, the material cause could not fulfil its role as the substratum persisting through generation and destruction: either the material cause would already have the formal cause, and thus could not gain it, or the material cause would itself be destroyed in the process of losing that formal cause and thus could not persist. There is, then, a fundamental asymmetry between the formal and material causes of perceptible objects: the material cause is necessary for the actualization of the formal cause, but the formal cause is accidental to the material cause. A formal cause cannot be part of the substratum persisting through the change in which that formal cause itself comes into being.

While the formal cause is accidental to the material cause, it is essential to the functional parts of perceptible objects. These functional parts, however, are part of the composite object, not of its material cause. The soul, for example, is essential to the organs of living beings, but the organs of living beings are part of something alive, and to be alive, in Aristotle's account, something must already have both a soul and a body, both a formal and a material cause. Thus, the organs of living beings are part of the composite object, the living organism, not of the non-living matter from which living organisms are made. In general, whenever a formal cause is essential to a part of a perceptible object, that part is a functional part and not a material cause, for functional parts cannot perform the principal job of the material cause: because they are dependent for their own nature upon the formal cause of the object to which they belong, they cannot persist through the processes in which that formal cause is gained or lost, namely generation and destruction.

Contrary to the traditional view, then, whatever acts as a material cause must be separable in time and definition from the formal cause to which it is joined to produce a composite object. The formal cause is essential to the composite object but accidental to the material cause of that composite object. The formal cause makes the composite object, but not its material cause, to be what it is. Composite objects have a second nature, grounded in their material cause.

Whatever else is required for this second nature, the material cause of a perceptible object must be something physical, for everything that perceptible objects do presupposes locomotion, and anything that undergoes locomotion must ultimately be made out of matter. Aristotle's position is not just that anything subject to locomotion must be a physical object. In addition, he argues that movable objects have to be made out of something physical; their material cause must be something extended, movable, and corporeal in its own right. Thus, regardless of their formal cause, perceptible objects must have a physical material cause, and their material cause will be something physical only if, in the end, it is made out of matter. Matter is the fundamental material cause of perceptible objects.

The independent nature of the material cause does not eliminate the possibility of essential change, for in generation and destruction neither the formal nor the material cause is generated or destroyed; the composite is. Thus, the mark of essential change is not the pure potentiality of the substratum, but the gain or loss of a formal cause. Even essential change requires something that persists throughout the change and remains upon its completion. Thus, the material cause persisting through an essential change must be independent of the formal cause gained or lost in that change. An underlying subject does not gain its being from its accidental properties, and the formal cause is accidental to the material cause.

10.2 The Material Cause and Potentiality

The material cause of a perceptible object, then, must have an actual, determinate set of properties of its own. In particular, it is responsible for the physical properties required to actualize the formal cause of a perceptible object. To be sure, Aristotle does say that the material cause is indeterminate.[1] This indeterminacy, however, is an external, relational property, belonging to a material cause by virtue of its relation to the object made from it, and is consistent with the material cause having an independent nature of its own. The material cause is indeterminate because it lacks certain properties that are crucial to the composite object made from it; these additional properties are supplied by the composite object's formal cause. The material cause, however, must be distinguished from the properties that it lacks; the absence of certain properties is over and above whatever the thing acting as a material cause is in its own right.[2] The material cause is indeterminate in relation to the determinate properties that can be added to it, not by virtue of its own nature.

The material cause is distinct from its indeterminacy in the same way that a substratum of change is distinct from its privation.[3] In the case of the bronze to be made into a statue, for example, what it is as bronze is distinct from what it also requires to be a statue. The privation belonging to a substratum is always qualified and restricted: it is the absence of some particular attribute or property. Because it is only the privation that is lost when the formal cause is acquired, the substratum remains as it was before. Whatever acts as a substratum of essential change is known as a substratum only in relation to the formal cause that is gained or lost in that change, but this relation to the formal cause cannot make the persisting substratum to be what it is.

Thus, whatever acts as a material substratum cannot be defined by its indeterminacy; this would turn every material cause into the opposite of its formal cause. Since opposites are mutually exclusive, this opposition would make an object composed of a material and formal cause impossible. Strictly speaking, a material cause is indeterminate, not because it is opposed to the kind of determination conferred by a formal cause, but because that determination is not already part of its own nature. The material cause is indeterminate only in relation to the formal cause that can be added to it or taken away. This indeterminacy, however, must be over and above what the material cause already is in its own right.

The four sublunary elements provide a good example of this qualified indeterminacy. Aristotle says that these four elements are potentially substances only because they are merely heaps.[4] The potentiality of the elements to be substances, however, does not entail that they do not exist, for clearly they do, even if only as heaps. Rather, the actuality missing in them is a particular kind of reality, namely the arrangement of mutually dependent parts that only organized wholes have, and heaps lack.[5] Thus, the indeterminacy that Aristotle attributes to these four elements must be understood in relation to the actuality of substances that are organized wholes and indivisible units. This qualified indeterminacy, however, does not strip the elements of their own actuality. Not only do they exist on their own, but they are also subject to generation and corruption. Thus, they must be composites of a material and a formal cause in their own right. Moreover, the four sublunary elements all possess natural motions of their own, as well as the basic tactile properties by which physical bodies interact. All of these characteristics are independent of the actuality, or determinate nature, of the more complex objects made out of the elements. The indeterminacy of these elements, then, is a qualified and restricted form of privation; they lack something in particular.

The same restricted privation is found in prime matter, the material cause of the four sublunary elements. The defining properties of these elements presuppose an extended, divisible, and movable substratum, one that persists through their generation and destruction. It is true that in Aristotle's view no perceptible object can be found without the basic tactile properties and natural motions of the elements. Thus, the corporeal substratum underlying the four elements is inseparable from these elemental properties in the sense that it never occurs apart from them. None of these elemental properties, however, makes their common substratum to be an extended, movable body in the first place. On the contrary, all of the specific, differentiating properties of the elements presuppose a physical substratum. Indeed, the nature of prime matter is the nature of matter *tout court*, because prime matter has whatever is left to matter when all of the properties are taken away that differentiate perceptible objects from one another. Thus, the inseparability of prime matter from these differentiating properties is consistent with it having an independent nature of its own, and this nature cannot be pure potentiality.[6]

The crucial point here is that the nature of a material cause cannot be grounded in the formal cause of the composite object made from that material cause. Of course, to be a material cause is to stand in a certain relation to some formal cause or other; in this sense, the material cause is dependent upon the formal cause. To say, however, that something is a material cause or a substratum is to give it a job to do, and whatever does this job must have an independent nature of its own. Moreover, because the being that the material cause lacks is a qualified one, the actuality conferred on a composite object by that object's formal cause is also a qualified kind of being, not being simply. The formal cause adds to the actuality already found in the material cause. This added actuality does not exclude any other kind of being in the material cause, except, of course, the privation of the formal cause. In sum, despite Aristotle's repeated association of the material cause with potentiality, the potentiality of the material cause of a perceptible object always presupposes a prior, independent nature in that material cause.[7]

In addition to its potentiality, a material cause is also passive inasmuch as it receives a formal cause.[8] In order to do this, however, a material cause must already possess certain properties and causal capacities in its own right. Thus, the privation, indeterminacy, passivity, and potentiality of the material cause must always be balanced against the hypothetical necessity that holds between material and formal causes. The material cause of every composite perceptible object must already possess certain properties in its own right if that composite object is to come into being and be able to do certain things. In addition, the material causes of perceptible objects must have the properties that all things subject to change of any kind require in their material cause. Taken together, these material properties are insufficient to explain the distinctive, ordered kinds of change that belong to composite perceptible objects. Without the right kind of material cause, however, there can be no motion or change in the first place. Thus, the material cause of a perceptible object cannot be understood simply as a principle of privation.

10.3 The Material Cause and Definition

Aristotle argues that perceptible objects are defined by their formal cause.[9] Individual perceptible objects are composite beings that can be destroyed by taking away their material cause, but, at least initially, their material cause is not included in their definition; composite objects are not indifferent to their material cause but are not defined by it. Aristotle, however, does not reduce perceptible objects to just their formal cause: he continues to describe them as composites of a material and formal cause, where the material cause is something distinct from the formal cause. If the definition of a perceptible object is, in the first instance, a statement of what its formal cause is, there must be more to a perceptible object than what is included in its definition.

In fact, when Aristotle describes more precisely the way in which the formal cause defines a composite perceptible object, it becomes clear that the material cause must have a discrete nature of its own.[10] The formal cause tells us what something is, but it does this by telling us what has been added to the material cause of the object in question. In effect, in asking what kind of thing a perceptible object is, we are really asking why it has the sorts of features it does, over and above what already belongs to it by virtue of its material cause. Thus, it is not the case that all of the permanent features of perceptible objects are due to their formal cause. Some of the material properties of perceptible objects are accidental to them in the sense of being merely local and transitory. Not all material properties, however, are accidental. On the contrary, as we saw in the previous chapters, some of the permanent features of perceptible objects belong to them because of their material cause.

It also follows that the relation between a formal and material cause is not an explanatory relation: a certain kind of material cause is needed in order to actualize a formal cause, but the formal cause does not explain the nature of that material cause. Otherwise, one and the same attribute will be explained in many different ways. If, for example, the natural motion of a perceptible object to a certain place in the cosmos is attributed to its formal cause, there will be many different explanations for the same motion, namely the many different formal causes possessed by the various perceptible objects in which this motion occurs. Aristotle, however, thinks these kinds of natural motion have the same explanation wherever they occur: they are due to the material element from which a perceptible object is predominantly made. The requirement for a certain kind of material cause explains why an object has the kind of material cause that it does; that requirement does not explain why that material cause has its distinctive properties and causal capacities in the first place. If anything, the requirement for a certain kind of material cause explains something about the formal cause and the object it defines, not the material cause of that object.

This view has implications for the way in which perceptible objects are defined. As we saw above, Aristotle says that, at least initially, the material cause is not part of the definition of a perceptible object. Since perceptible objects possess certain necessary properties and causal capacities by virtue of their material cause, their definition does not capture all of their necessary properties and causal capacities. If the essence of a perceptible object consists in what is captured by its formal cause, perceptible objects have necessary attributes that are not essential attributes. The formal cause may specify what kind of thing a perceptible object is, but the formal cause does not include everything necessary to that object.[11] Thus, it is not the case that whatever falls outside the formal cause of a perceptible object is merely accidental to that object. On the contrary, many material attributes are permanent features of the perceptible objects to which they belong, and anything permanent cannot be accidental.[12]

Restricting the range of what is explained by a perceptible object's formal cause does not take away its priority with respect to defining perceptible objects. On the contrary, the hypothetical necessity that holds between perceptible objects and their raw materials preserves the priority of formal causes over material causes without making material causes dependent for their own being on the formal cause of the object made from them. Most plants require light and water to live, but water and light are not defined in relation to the growth of plants. In the same way, the raw materials of a perceptible object are necessary for the generation and proper functioning of that object, but these raw materials retain their own properties and causal capacities.

Some commentators argue that if there is an underlying material cause with a distinct, permanent nature of its own, the formal cause will be reduced to a set of properties inhering accidentally in the material cause.[13] This objection fails, however, because it gives the formal cause too much to do: the formal cause is now responsible for the properties and causal capacities of the material cause as well. For the reasons given above, the formal cause cannot define a composite object's material cause. The formal cause may define the composite object of which the material cause is a part, but it does this by specifying what has been added to the material cause in order to constitute the composite object. The formal cause does not need to be essential to the material cause; it needs only to be essential to the composite object.

10.4 The Material Cause and Change

Aristotle, then, has a concept of matter as physical stuff, which he arrives at by means of an analysis of the material causes of perceptible objects and the functions they perform. One result of this analysis is that the connection between matter and the material cause is not accidental. While not every material cause is something physical, the most characteristic function of a material cause is to be a substratum of generation and destruction, and only matter or something made out of matter can act as such a substratum, for generation and destruction require efficient causes, and efficient causality in perceptible objects always involves physical contact between extended, movable bodies; as we saw in chapter 3, that contact requires a common, physical substratum in the affected bodies. The physical characteristics required by a substratum of generation and destruction must be present throughout these processes and so cannot be due to the formal cause that is being gained or lost. Whatever else they are, the material causes that underlie generation and destruction must be something physical. The most characteristic function of the material cause can be exercised only by matter.

The systematic connection between matter and the material cause is also found in the other forms of change that perceptible objects undergo. As we saw in section 1.1.3, everything that perceptible objects do presupposes locomotion, and locomotion takes place only in physical objects, for physical objects cannot be in several places at once; at any given moment, they have a unique location.[14] Only movable bodies, however, have a unique location and only physical bodies are movable.[15] Thus, if something is capable of locomotion, it must have a physical body. Physical bodies, in turn, must ultimately be made out of matter. In addition to giving perceptible objects a particular location, matter individuates numerically discrete substances that share the same formal cause; only a material cause with a physical nature of its own is able to do this. In sum, matter is the paradigmatic instance of a material cause because of its foundational role in locomotion and, thereby, in change of all types.

The claim that the material causes of perceptible objects must have properties and causal capacities of their own is entirely compatible with Aristotle's insistence that perceptible objects are not reducible to their material causes. These objects, he argues, must have a formal cause over and above their material cause, because their distinctive behaviour cannot be explained solely by the physical characteristics of their raw materials. Something more is required to explain the ordered patterns of change in the physical universe beyond the matter from which perceptible objects are made. This deficiency in matter, however, is consistent with the view that perceptible objects need a material cause with a physical nature of its own. The formal and material causes of perceptible objects are mutually irreducible, because each contributes something necessary to the object put together from them, something that the other lacks. Neither is sufficient to constitute a perceptible object. The insufficiency of matter to constitute perceptible objects, however, does not entail that the material causes of perceptible objects are accidental to these objects or purely potential. On the contrary, the specific capacities that belong to perceptible objects presuppose lower-level capacities in their material causes.

The upshot is that perceptible objects typically have multiple natures. These natures begin with their lowest-level material substratum and are increased through the addition of formal causes. Given the multiple natures of perceptible objects, explaining their behaviour with reference to the correct level of composition is as important as identifying the correct causal agent. In particular, an explanation is incorrect if it attributes to a formal cause a causal capacity that belongs to a lower-level material cause; as we saw, motion to a natural place is correctly explained by the elements from which a moving object is made, not by that object's formal cause. The physical necessity governing the behaviour of the material elements operates independently of the formal causes added to them in more complex perceptible objects. An explanation of these necessary physical features that appeals to the higher-level formal causes of perceptible objects occurs at the wrong level. In this sense, these explanations are category mistakes.

Some explanations are incorrect because they begin at too high a level. Others are incorrect because they start too low. Since teleology among perceptible objects is found only in the changes of instrumentally organized objects or groups of objects, the good operates only above the level of the physical necessity governing the material elements. With respect to the behaviour of perceptible objects, the job of formal causes is to explain certain types of *ordered* change, not the possibility of change as such. Formal causes make perceptible objects capable of more ordered kinds of change by adding causal capacities over and above the ones that perceptible objects already have by virtue of their raw materials. In this sense, formal causes make perceptible objects more powerful; the latter can now do more. Formal causes, however, do not make perceptible objects more powerful in the sense of adding to the causal powers required to set bodies in motion or change them; the latter belong to the raw materials of perceptible objects and are not increased by the addition of further levels of organization. Formal causes make the motions and changes produced by perceptible objects more ordered and complex; they do not increase the total amount of motion and change that these objects can produce.

Underlying this view of the material causes of perceptible objects is Aristotle's claim that all changeable objects are *composite* beings.[16] This composite character is seen already in Aristotle's view that everything undergoing a change must both stay the same and become different.[17] A changing object must stay the same because change is not a process in which one object is simply replaced by another; it must also become different, for otherwise no change takes place. In order to satisfy these requirements, Aristotle concludes that anything undergoing a change of any kind must be one in number, but two in kind.[18] Every changeable object is made up of, at the very least, a persisting substratum and one of a pair of contraries, and the substratum and the contrary present in it are different kinds of being. Even though the substratum and contraries are inseparable in that there is no substratum of a changeable object without some contrary or other, changeable objects must still be understood as composite things, for part of them persists, and that part, the substratum, must be independent of the part that does not persist, the contraries.

The composite character of perceptible objects is also seen in their causal capacities: these occur in an ordered sequence in which the higher-level ones presuppose the lower-level ones. The higher-level causal capacities result from the addition of a formal cause to a pre-existing material cause, thereby adding a second set of causal capacities, over and above the ones that already belong to the raw materials. In the case of physical human artefacts, these additional causal capacities depend upon human agency for both their occurrence and use; in the case of natural substances, they occur and are used without human agency. In both cases, these additional causal capacities and the ends to which they are directed belong to the composite perceptible object alone; neither its material nor its formal cause can exercise these causal capacities on its own. The composite character of perceptible objects, then, is also seen in their efficient and final causes: their specific causal powers and ends are found only in suitably constituted composites and are lost if either the formal or the material cause of these composites is missing.

The traditional view of Aristotle's material cause does see perceptible objects as composites of a formal and material cause. It renders this composite nature nugatory, however, by making the material cause dependent for its own attributes upon the formal cause of the object made from it. This radical compositeness is lost if the material cause has no attributes of its own, and, as such, has nothing to contribute to the object made from it. Similarly, the unity of natural substances should not be secured by making their formal causes responsible for all of their necessary attributes. If, for metaphysical reasons, natural substances are held to lack a second set of properties and causal capacities, grounded in their material cause, Aristotle's metaphysics becomes inconsistent with his natural philosophy; to pick just one example, this view makes generation and destruction impossible. This inconsistency is not alleviated by asserting the greater authority of Aristotle's metaphysics over his natural philosophy. Aristotle's science of being qua being discovers axioms that all of the particular sciences must follow, such as the principle of non-contradiction. Aristotle, however, never assigns to this universal science the determination of the principles of the particular sciences.[19] Thus, the science of being qua being is not responsible for the principles of change; that is the job of the science of change. The science of being qua being also cannot contradict the principles of the latter. If the principles of Aristotle's ontology are allowed to violate his principles of generation and change, the result will be an ontology that populates the perceptible universe with generated substances that, according to his natural philosophy, could never have come into being in the first place.

It is not necessary, however, to eliminate the independence of the material cause in order to preserve the unity of perceptible objects, for one of the fundamental principles of Aristotle's natural science is that causal powers can be added without adding things. The addition of causal capacities through successive levels of organization does not compromise the numerical unity of individual perceptible objects. One thing can have many natures. The unity of perceptible objects does not require the simplicity of their nature. The material cause of a perceptible object must have a nature of its own, even if that nature is insufficient to constitute the entire being of the perceptible object made from it. Thus, far from everything permanent about perceptible objects being due to their formal cause, it is never the case that the formal cause of a perceptible object explains all of that object's permanent features. Perceptible objects, including natural substances, are radically composite beings, with distinct material and formal natures.

Notes

Preface

1 For Darwin's view of Aristotle's biology, see A. Gotthelf, "Darwin on Aristotle," *Journal of the History of Biology* 32 (1999): 3–30. For a favourable assessment of Aristotle's biology by a modern biologist, see Armand Leroi, *The Lagoon: How Aristotle Invented Science* (New York: Viking, 2014).

2 For example, S. Sambursky, *The Physical World of the Greeks* (London: Routledge & Paul, 1956; Princeton: Princeton University Press, 1987), 97, writes that Aristotle's influence on physics was retrograde, because he reversed the progress that had been made before him by the atomists. Thomas Kuhn takes a stronger view and argues that Aristotle's physics is wholly incompatible with modern Newtonian physics: "Preface," *The Essential Tension*, xi–xii (Chicago: University of Chicago Press, 1977); and Kuhn, "What Are Scientific Revolutions?" in *The Probabilistic Revolution*, vol. 1, *Ideas in History*, ed. L. Kruger, L.J. Daston, and M. Heidelberger, 7–22 (Cambridge, MA: MIT Press, 1987), reprinted in Kuhn's *The Road since Structure*, ed. J. Conant and J. Haugeland, 13–32 (Chicago: University of Chicago Press, 2000).

Introduction

1 *APo.* I 2, 71b9–12; II 2, 90a6–8; *Ph.* I 1, 184a10–15; II 3, 194b17–20; *Metaph.* I 1, 981a24–b2.

2 *APo.* I 2, 72a14–18; 6, 74b25–6, 75a28–31; 7, 75a38–b20; 9, 75b37–76a15; 10, 76a37–b16; 28, 87a38–b4.

3 *Cael.* IV 3, 310a27–32; *GA* II 6, 743a18–27.

4 W. Jaeger, *Aristoteles* (Berlin: Weidmann, 1923), 408; A. Mansion, *Introduction à la physique aristotélicienne*, 2nd ed. (Louvain: Institut supérieur de

philosophie, 1946), 154–5; W.D. Ross, *Aristotle*, 5th ed. (London: Methuen, 1949), 66, 69, and 73; F. Solmsen, "Aristotle and Prime Matter: A Reply to Hugh R. King," *Journal of the History of Ideas* 19 (1958): 243–52; and Solmsen, "Aristotle's Word for 'Matter,'" in *Didascaliae*, ed. S. Prete, 393–408 (New York: B. Rosenthal, 1961); M. Grene, *A Portrait of Aristotle* (London: Faber & Faber, 1963), 196 and 200; W. Wieland, *Die aristotelische Physik*, 2nd ed. (Göttingen: Vandenhoeck & Ruprecht, 1970), 140n29, 173–87, and 209–11; C.J.F. Williams, *Aristotle's De generatione et corruptione* (Oxford: Clarendon, 1982), 211–19; K. Fine, "Aristotle on Matter," *Mind* 101 (1992): 37–57.

5 T. Kuhn, "What Are Scientific Revolutions?" in *The Road since Structure* (2000), 17, writes, "In Aristotle's physics ... matter is very nearly dispensable." S. Waterlow, *Nature, Change, and Agency in Aristotle's* Physics (Oxford: Clarendon, 1982), 35, writes, "In mechanics ... there is no place for the concept of specifically different substances. All bodies can be shown to behave at all times in accordance with the same set of laws, because the laws depend on the properties of body as such. These properties, moreover, according to Aristotle's categorical scheme, fall into categories other than that of Substance: mass, velocity, position, duration, etc. The same individual Aristotelian substance could alter in respect of all these without impairment to its substantial nature. Thus mechanics studies, through their effects, the 'natures' of these properties in their determinations and combinations. Hence in a mechanist account of an object's behaviour, none of that behaviour can appear as issuing from a specific substantial nature." She goes on to argue that because Aristotle is interested only in the specific natures of natural substances, he has no room for mechanics; for him, the properties studied by mechanics are accidental to natural substances.

6 *Ph.* I 9, 192a31; II 3, 194b23–195a26; *GC* I 4, 320a2–3; *Metaph.* VII 7, 1032a17.

7 The classic statement of this position is found in Aquinas: *Questiones disputatae de potentia*, q. 4, a. 1, response; *Questiones disputatae de anima*, q. 1, a. 9, response; *Summa theologiae* I, q. 66, a. 1, response. More recent commentators who defend this position include J. Owens, "Matter and Predication in Aristotle," in McMullin, *The Concept of Matter*, 99–115, esp. 112, reprinted in his *Aristotle: The Collected Papers of Joseph Owens*, ed. J. Catan, 35–47 (Albany, NY: SUNY Press, 1981); W. Charlton, *Aristotle's Physics I & II* (Oxford: Clarendon, 1970), 75–7; J.-M. LeBlond, "Aristotle on Definition," in Barnes, Schofield, and Sorabji, *Articles on Aristotle*, 3:63–79; L. Gerson, "Artifacts, Substances, and Essences," *Apeiron* 18 (1984), 50–8; S. Cohen, "Aristotle on Heat, Cold, and Teleological Explanation," *Ancient Philosophy* 9 (1989), 255–70; E. Katayama, *Aristotle on Artifacts: A Metaphysical Puzzle* (Albany, NY: SUNY Press, 1999); C. Frey, "Organic

Unity and the Matter of Man," *Oxford Studies in Ancient Philosophy* 32 (2007), 167–204; M. Scharle, "Material and Efficient Causes in Aristotle's Natural Teleology," in *Aristotle on Life*, ed. J. Mouracade, *Apeiron* 41, no. 4 (2008), 27–45. Some commentators restrict this claim to living natural substances: L.A. Kosman, "Animals and Other Beings in Aristotle," in *Philosophical Issues in Aristotle's Biology*, ed. A. Gotthelf and J. Lennox, 360–91 (Cambridge: Cambridge University Press 1987); M.L. Gill, *Aristotle on Substance: The Paradox of Unity* (Princeton: Princeton University Press, 1989), 146–67.

8 *GA* IV 3, 769b12; 4, 770b16–18. The malformation Aristotle is considering in these passages is the relatively infrequent occurrence of deformed animals within a species. In addition, he considers whole species to be deformed when all of their members have dysfunctional organs, e.g., the blind eyes of moles (*HA* I 9, 491b27–36; IV 8, 532b34–533a12), or biological capacities that do not function as well in one species as in others, e.g., aquatic animals such as seals that are not very good at walking on land, as compared to terrestrial quadrupeds (*HA* II 1, 498a33–b1; *PA* II 12, 657a22 –24; IV13, 697b1–8). This second kind of deformation is different from the first because it belongs to whole species and is thus regular and frequent. Still, the material cause is part of Aristotle's explanation of these deformities as well. On deformed species, see H. Granger, "Deformed Kinds and the Fixity of Species," *Classical Quarterly* 37 (1987): 110–16; and C. Witt, "Aristotle on Deformed Kinds," *Oxford Studies in Ancient Philosophy* 43 (2012), 83–106.

9 E. Hartman, *Substance, Body, and Soul: Aristotelian Investigations* (Princeton: Princeton University Press, 1977), 54, writes, "Regularity is the province of forms, spontaneity of matter."

10 R. Westfall, *The Construction of Modern Science: Mechanisms and Mechanics* (Cambridge: Cambridge University Press, 1977), 19, writes, "To Aristotle, motion had been a process involving the very essence of a body, a process whereby its being was enhanced and fulfilled."

11 Aquinas, *Super libros De generatione et corruptione*, bk I, lect. 10, #80; McMullin, "Introduction," in McMullin, *The Concept of Matter*, 7; and J.J. FitzGerald, "'Matter' in Nature and the Knowledge of Nature: Aristotle and the Aristotelian Tradition," in McMullin, *The Concept of Matter*, 79–98, esp. 86. Also, Cohen, "Aristotle on Heat, Cold, and Teleological Explanation," 255–70.

12 For the Thomistic defence of the traditional view, see Aquinas, *De potentia*, q. 4, a. 1, response; *De anima*, q. 1, a. 9, response; *Summa theologiae* I, q. 66, a. 1, response; *Super libros De generatione et corruptione*

I, lect. 10, #80; FitzGerald, "'Matter' in Nature and the Knowledge of Nature," 79–98; Owens, "Matter and Predication in Aristotle," 99–115; and N. Luyten, "Matter as Potency," in McMullin, *The Concept of Matter*, 122–33. See also Carl Page, "Predicating Forms of Matter in Aristotle's *Metaphysics*," *Review of Metaphysics* 39 (1985): 57–82. For non-Thomistic defenders of the traditional view, see Solmsen, "Aristotle and Prime Matter"; H.M. Robinson, "Prime Matter in Aristotle," *Phronesis* 19 (1974), 168–88; P. Suppes, "Aristotle's Concept of Matter and Its Relation to Modern Concepts of Matter," *Synthese* 28 (1974): 27–50; A. Code, "The Persistence of Aristotelian Matter," *Philosophical Studies* 29 (1976): 357–67; J. Kung, "Can Substance Be Predicated of Matter?" *Archiv für Geschichte der Philosophie* 60 (1978): 140–59; Williams, *Aristotle's De generatione et corruptione*, 211–19; Fine, "Aristotle on Matter." The most succinct statement of the traditional view is found in Robinson, "Prime Matter in Aristotle," 167–8 and 187.

13 M.F. Burnyeat, "Is an Aristotelian Philosophy of Mind Still Credible?" in Nussbaum and Rorty, *Essays on Aristotle's De anima*, 15–26, writes, "Aristotle's philosophy of mind is no longer credible because Aristotelian physics is no longer credible" (16), and "Aristotle has what is for us a deeply alien conception of the physical" because he holds that the bodies of animals are essentially alive (26). See also J.L. Ackrill, "Aristotle's Definitions of *psuche*," in Barnes, Schofield, and Sorabji, *Articles on Aristotle*, 4:65–75.

14 A. Code and J. Moravcsik, "Explaining Various Forms of Living," in Nussbaum and Rorty, *Essays on Aristotle's De anima*, 129–45, esp. 143, argue, "For Aristotle the concept of matter is species-relative, and should itself be seen as including a teleological aspect – specific potentialities defined in terms of their exercise." Sambursky, *The Physical World of the Greeks*, 81, blames Aristotle's negative influence on physics on his reliance on teleology.

15 M. Frede, "On Aristotle's Conception of the Soul," in Nussbaum and Rorty, *Essays on Aristotle's De anima*, 93–107, esp. 100. R.J. Hankinson, *Cause and Explanation in Ancient Greek Thought* (New York: Oxford University Press, 1998), 130, writes, "Matter is, for Aristotle, a relational concept … and there is no such thing as matter as such: matter is the matter *for* something."

16 Code and Moravcsik, "Explaining Various Forms of Living," 139.

17 Other commentators who hold that the basic properties of natural substances are not exhausted by their formal cause include: R. Boehm, *Das Grundlegende und das Wesentliche* (Den Haag: Nijhoff, 1965), esp. 8–26

and 40–54; R.E. Allen, "Substance & Predication in Aristotle's *Categories*," in *Exegesis and Argument*, edited by E.N. Lee, P.D. Mourelatos, and R.M. Rorty, supplement, *Phronesis* 1 (1973): 362–73; S. Mansion, "The Ontological Composition of Sensible Substances in Aristotle (*Metaphysics* Z 7–9)," in Barnes, Schofield, and Sorabji, *Articles on Aristotle*, 3:80–7; J. Cooper, "Hypothetical Necessity," *Aristotle on Nature and Living Things*, ed. A. Gotthelf, 151–67 (Pittsburgh: Mathesis Publications, 1985); T. Scaltsas, "Substratum, Subject, and Substance," *Ancient Philosophy* 5 (1985), 215–40; M. Furth, *Substance, Form and Psyche: An Aristotelian Metaphysics* (Cambridge: Cambridge University Press, 1988); M. Wedin, *Aristotle's Theory of Substance* (Oxford: Oxford University Press, 2002), esp. 427–41.

Chapter One

1 *de An.* II 5, 416b32–417a20, b19–28; 11, 424a1; 12, 424a21–32; III 1, 424b27–30; 12, 434b11–14 and 434b27–435a10; 13, 435a17–18.
2 *GC* I 6, 322b21–4, 323a10–12.
3 *GC* I 7, 324a24–b4. The example he uses here is of a doctor curing a patient: even if there is no direct contact between them, there is between the patient and the medicine that effects the cure.
4 *GC* I 7, 323b29–34.
5 At *de An.* III 7, 431a4–7, Aristotle says that a sense organ is not affected or altered when engaged in perception, but is brought from a state of potentiality to actuality in that it exercises its previously inactive capacity. Nevertheless, he wants to preserve his prior claim that in perception an external object moves and affects the perceiver. The way he does this is to introduce a special kind of alteration, namely the transition from having a capacity to exercising it, as distinct from the usual kind of alteration, which always involves the destruction of one state of affairs when its opposite comes to be. This special kind of alteration, however, still presupposes contact; as a result, a sense organ can be destroyed if the sense impression is too violent. See *de An.* II 5, 417a2–b16, 418a1–6; 11, 424a1–5; 12, 424a21–32; III 1, 424b27–30; 12, 434b11–14 and 434b27–435a10; 13, 435a11–19, b7–19; *Ph.* VII 2, 244b2–245a11.
6 *de An.* II 7, 419a13–21.
7 *Cael.* I 7, 275b5–7. At *Cael.* I 7, 275b7–12, Aristotle states that all bodies that occupy a place are perceptible, and that if something is not in a place, it is not a perceptible body. Taken together, these two statements imply that all and only perceptible bodies occupy a place.
8 *Cael.* II 7, 289a20–35; *Mete.* I 3, 341a13–32. This process is discussed in greater detail below in section 5.3.

9 *Cael.* I 2, 268b15–17; 9, 278b22–279a5; *Ph.* III 5, 205a10–12, b24–35; IV 4, 211a2–6.

10 *Ph.* III 1, 200b12–25; *Metaph.* XIII 3, 1077b22–7.

11 *Ph.* VI 4, 234b10–235a10, b1–5; VII 1, 242a40; VIII 6, 258b24–5; 10, 267a22–3; *Cael.* I 1, 268a6–7.

12 *Ph.* IV 5, 212b28–9; 4, 212a6–7.

13 *Ph.* IV 1, 208b22–5; *Cael.* III 6, 305a25–7; *Metaph.* XIV 5, 1092a19.

14 *Ph.* IV 1, 209a6–7; 6, 213b20; *GC* I 5, 321a7–9, b15–16; *de An.* II 7, 418b17.

15 *GC* I 2, 315b24–33; 5, 320b14–17, 22–8; II 1, 329a13–24; *Cael.* III 1, 299a2–12; 7, 306a23–30.

16 *Ph.* IV 1, 208a31–3; VII 2, 243a39–40; VIII 7, 260a26–261a28.

17 *Ph.* VIII 7, 260b11–13.

18 *Ph.* VIII 7, 261a1–7; *GC* II 10, 336a20–6; *Metaph.* VIII 1, 1042b2–6; XII 7, 1072b8–9, 1073a12.

19 *Ph.* I 7, 190a13–191a17; *Metaph.* VII 9, 1034b7–16.

20 *Ph.* I 8, 191b12–27; 9, 192a3–6, 29–34; *GC* I 3, 319a17–22; *Cael.* III 2, 301b32–302a6; *Metaph.* VII 8, 1033a24–b5.

21 *Metaph.* VI 1, 1026a2–4; VII 7, 1032a20–2; 8, 1033b18–19; 9, 1034b12; 10, 1035a25–7; 15, 1039b29–30; VIII 1, 1042a26; 5, 1044b27–9; XII 2, 1069b24–6; *GC* I 7, 324b4–6.

22 *Ph.* I 7, 190a13–21, b10–17, 23–5, 33–191a3. This claim is the centrepiece of his response to Parmenides: because every change has a persisting substratum, change never involves something coming to be from nothing or being reduced to nothing; because of the exchange of contraries, however, change does make the world different. See also *Ph.* I 8, 191a23–b27; 9, 192a29–34; *Metaph.* VII 8, 1033a24–b5.

23 For example, Waterlow, *Nature, Change, and Agency in Aristotle's* Physics, argues that Aristotle thinks about motion only in relation to natural motion: "In Book III Aristotle treats natural change as the type and model of change as such" (95); "his entire conception of change in Book III is governed by the logic of natural change" (99).

24 H. Butterfield, "The Historical Importance of a Theory of Impetus," in his *Origins of Modern Science*, rev. ed. (London: G. Bell, 1957), 15–16, argues that an essential feature of Aristotle's view of motion was the claim "that a body would keep in movement only so long as a mover was actually in contact with it, imparting motion to it all the time."

25 H. Carteron, "Does Aristotle have a Mechanics?" in Barnes, Schofield, and Sorabji, *Articles on Aristotle*, 1:161–74, 170n41, argues that Aristotle recognizes inertia in the sense of the resistance of physical objects to externally caused change but lacks the principle of inertia, because he

denies that a body could move indefinitely in space. R.J. Hankinson, "Chapter 5: Science," in *The Cambridge Companion to Aristotle* (Cambridge: Cambridge University Press 1995), 146n5, argues that Aristotle never sufficiently effects "the divorce between the notions of force and motion which the modern concept requires."

26 *Ph.* VII 1, 241b34; VIII 4, 254b24–33.

27 T. Kuhn argues that inertial motion is conceptually impossible for Aristotle, because Aristotle makes place a quality, and locomotion thus becomes a change of qualities: "Preface," xi–xii; Kuhn, "What Are Scientific Revolutions?" Aristotle, however, repeatedly says that a place must be separable from the physical body that occupies it, because it is possible to replace one body with another in the same place: *Ph.* IV 1, 208b7 and 28; 4, 210b34–211a3, 212a1. A place, then, is not a quality of a body, because qualities, unlike places, always move with the object to which they belong.

28 G. Sarton, *A History of Science* (Cambridge, MA: Harvard University Press, 1952), argues that, for Aristotle, "things change … in order to attain or to approach their perfection" (1:497), and "every motion has a direction and a purpose. The direction is toward something better or more beautiful" (1:515). M. Matthen, "The Four Causes in Aristotle's Embryology," in *Nature, Knowledge and Virtue, Apeiron* 22, no. 4 (1989): 159–79, argues that in Aristotle's physics teleology goes "all the way down" and "every cause is teleologically specified" (160, 162–3). Matthen does not discuss any of the passages where Aristotle says some things do *not* happen for the sake of a final cause; on this point, see chapter 9 below. In support of his claim, he cites *Ph.* V 1, 224b7–8, repeated at 5, 229a25–6: "It is that to which, rather than that from which, the motion proceeds that gives its name to change." Matthen interprets the "that to which" in this text to be a final cause, but the expression Aristotle uses here (εἰς ὅ) is not the one he uses to refer to the final cause, namely τὸ οὗ ἕνεκα. In fact, in the case of the destruction of a substance, the "that to which" it changes is precisely not its final cause (*Ph.* II 2, 194a29–33).

29 *Ph.* III 1, 200b12–25.

30 *Ph.* II 1, 192b8–19, 193a29–30; *Cael.* I 2, 268b16; III 2, 301b18–19; *de An.* II 1, 412b15–17; *GA* II 1, 735a2–5; *Metaph.* V 4, 1015a13–15; IX 8, 1049b5–10. This topic is considered in greater detail below in chapter 7.

31 *Metaph.* IX 8, 1049b12–15. Note that Aristotle's claim here is about efficient causes in general, not just ones acting for the sake of a final cause, even if teleological examples illustrate this point most clearly.

32 At *Ph.* V 6, 230a18–b10, Aristotle says that, among the different types of change, only locomotion is either according to nature or contrary

to nature, but many types of alteration and change of size are neither, inasmuch as one and the same change can sometimes be according to nature and sometimes contrary to nature; for example, becoming paler or darker, or even becoming healthy or sick. Even destruction can sometimes be according to nature: the death of an animal is according to nature when it occurs through old age, but contrary to nature when caused violently.

33 *Ph.* III I, 201a10–11. R. Heinaman, "Is Aristotle's Definition of Change Circular?" *Apeiron* 27 (1994): 25–37, argues correctly that Aristotle's definition of change is circular. Still, as E. Hussey argues in his *Aristotle's Physics Books III & IV* (Oxford: Clarendon, 1983), xiii, it is useful because it grounds change in other fundamental features of the world.

34 *Ph.* III I, 201b5–13; 2, 201b18–35.

35 *Ph.* I 7, 190a13–191a17; *Metaph.* VII 9, 1034b7–16.

36 *Ph.* III I, 201a29–b3.

37 *Ph.* III I, 200b32–201a15.

38 *Ph.* II 3, 195a23–6; 7, 198b8–9; 8, 198b17; *GA* II 1, 731b23–4; V 8, 789b5–6; *Metaph.* V 2, 1013b25–8; *APo.* II 11, 95a6–8.

39 *Ph.* II 2, 194a30–3.

40 R. Sorabji, *Matter, Space, and Motion: Theories in Antiquity and Their Sequel* (Ithaca, NY: Cornell University Press, 1988), 187, points out that Aristotle's definition of place as a surrounding surface and his definition of natural place are distinct, but often occur next to one another in his texts, e.g., *Cael.* IV 3, 310b7–11; *Phys.* IV 1, 208b1–8, 8–11, 11–22; 4, 211a3–6; 5, 212b29–213a 10; 8, 214b12; 215a8–11. B. Morison, *On location: Aristotle's Concept of Place* (Oxford: Clarendon, 2002), 20–5, makes clear that Aristotle's account of place comes before, and is independent of, his account of natural place; the former is based on the displacement and replacement of bodies in a location, the second on the natural motion of the elements.

41 *Ph.* IV 2, 209b28–33; 4, 212a2–6.

42 *Ph.* IV 2, 209b22–30; 3, 210b27–30; 4, 212a14–16.

43 *Ph.* IV 2, 209a35–b2; 4, 211a1–2, 27–33; 212a14–18.

44 *Ph.* IV 4, 212a19–21. In the case of bodies moving in a circle, where there are no external bodies beyond them, e.g., the outermost sphere of the universe, they move in relation to one another, that is, the various parts of the moving sphere or circle surround one another: *Ph.* IV 5, 212a31–b1.

45 *Ph.* IV 4, 212a16–20.

46 *Ph.* IV 4, 211a17–23.

47 *Ph.* IV 4, 211a29–b5.

48 *Ph.* IV 4, 211b25–9.

49 *Ph.* IV 4, 212a20–4.

50 *Cael.* II 13, 293b11–15.
51 *Ph.* IV 4, 212a24–9; b29–34.
52 For the mechanism by which the natural motions of the elements take place, see chapter 5 below.
53 *Mete.* I 4, 342a28–33.
54 Thus, some authors have argued that a statement of the principle of inertia can be found in Aristotle's *Physics*: T. Heath, *Mathematics in Aristotle* (Oxford: Clarendon, 1949), 115–16; Furth, *Substance, Form and Psyche*, 69. According to I.B. Cohen, Newton also thought that a statement of the principle of inertia could be found in Aristotle, as well as in the ancient atomists: "Quantum in se est: Newton's Concept of Inertia in Relation to Descartes and Lucretius," *Notes and Records of the Royal Society of London* 19 (1964): 131–55, esp. 140–1.
55 *Ph.* IV 8, 215a19–22; *Cael.* III 2, 301b2–4.
56 *Ph.* IV 8, 215a1–14; *Cael.* III 2, 300b9–17.
57 *Ph.* IV 8, 215a14–19.
58 *Cael.* III 2, 301b17–30; *Ph.* VIII 10, 266b29–267a11.
59 *Cael.* I 2, 269a1–2; IV 4, 311a30–b8; *GC* II 8, 334b31–4.
60 *Cael.* I 2, 268b27–269a2; 3, 269b34.
61 *Cael.* III 2, 301b20.
62 *Ph.* II 1, 192b8–27.
63 On the role of contact in physical causation, see chapter 2 below.

Chapter Two

1 A number of commentators have made this point; a good recent example is Hankinson, *Cause and Explanation in Ancient Greek Thought*, 125.
2 *Ph.* V 1, 224b5–8; *GC* I 7, 324b14–18; II 9, 335b7–24; *de An.* I 1, 403a24–b19; *GA* II 6, 743a21–7; *Metaph.* I 9, 991a8–11; 992a25; VII 8, 1033b26–1034a8; VIII 3, 1043b16–18; XII 6, 1071b14–16.
3 *Ph.* II 1, 193a9–b8; 2, 194a18–28; 8, 198b10–16; *GC* II 6, 333b3–20; 9, 335b24–35; *PA* I 1, 640a20–7, b5–29, 641a7–18; *Metaph.* I 8, 988b26–9, 989a25–6; XII 6, 1071b28–37.
4 *Ph.* II 3, 194b29–32, 195a21–3, 30–36, b5–6; III 2, 202a9–12; *GC* I 7, 324a8–14; *Metaph.* VII 9, 1034b16–19.
5 *de An.* I 4, 408b11–15; *GC* II 9, 335b20–9.
6 *Ph.* II 2, 194b13; 7, 198a24–7; III 2, 202a9–12; *GC* I 5, 320b17–21; *Metaph.* VII 7, 1032a25; 8, 1033b33; 9, 1034b17; IX 8, 1049b24–7; XII 3, 1070a8, 27–8, b34; XIV 5, 1092a16.
7 *de An.* I 1, 403a24–b19; 4, 408b11–15; b30–1; III 4, 429b13–14.

8 *Ph.* V 1, 224b5–8; *GC* I 7, 324b14–18; II 9, 335b18–24; *Metaph.* I 9, 991a8–11; 992a25; VII 7, 1032a12–25; 8, 1033b26–1034a8; VIII 3, 1043b16–18; XII 6, 1071b14–16.

9 *GC* I 6, 322b11–21; 10, 328a19–22.

10 *Ph.* III 2, 202a3–9; VII 1, 242b59–63; 2, 243a34–5; *GA* II 1, 734a3–5; 4, 740b22–741a4.

11 *de An.* I 1, 403a24–b12; 4, 408b11–15; III 4, 429b13–14.

12 *Cael.* IV 3, 310a27–32; *GA* II 6, 743a18–27.

13 *Metaph.* VII 17, 1041a6–22.

14 *Ph.* III 2, 202a9–12; *de An.* II 5, 417a17–18; *GA* II 6, 743a18–27.

15 Hankinson, *Cause and Explanation in Ancient Greek Thought*, 411 and 454, calls this the Principle of Prior Actuality, and again argues that it is older than Aristotle.

16 *Ph.* II 2, 194b13; 7, 198a24–7; III 2, 202a9–12; *GC* I 5, 320b17–21; *Metaph.* VII 7, 1032a25; 8, 1033b33; 9, 1034b17; IX 8, 1049b24–7; XII 3, 1070a8, 27–8, b34; XIV 5, 1092a16.

17 *Metaph.* VII 9, 1034a33–b7; *GA* III 11, 762a18–27. For a discussion of Aristotle's views on spontaneous generation, see J. Lennox, "Teleology, Chance, and Aristotle's Theory of Spontaneous Generation," *Journal of the History of Philosophy* 20 (1982), 219–38; and A. Gotthelf, "Teleology and Spontaneous Generation in Aristotle: A Discussion," in *Nature, Knowledge and Virtue*, *Apeiron* 22, no. 4 (1989), 181–93. Despite their differences on the implications of spontaneous generation for teleology, Lennox and Gotthelf agree that the efficient cause of spontaneous generation – typically the heat in the environment – has to bring about the right sorts of changes in the appropriate raw materials in order to generate the resulting biological organism. Thus, spontaneous generation is not uncaused generation, and something is transmitted from the causal agent to the raw materials.

18 *GC* I 5, 320b17–21.

19 *Cael.* IV 3, 310a27–32; *GA* II 6, 743a18–27.

20 *Ph.* II 3, 194b35–195a1, 5–8, b23–4; III 2, 202a9–12; *de An.* III 7, 431a3; *Metaph.* IX 8, 1049b24–7.

21 *GC* I 7, 323b3–324a24.

22 *de An.* II 5, 417a17–18; *GA* II 1, 734a30–3; *Metaph.* VII 7, 1032a24; 9, 1034a21–5.

23 *Ph.* II 1, 193a28–b12; 2, 194a12–28; 8, 198b10–16; *GC* II 6, 333b3–20; 9, 335b24–35; *PA* I 1, 640a20–7, b5–29, 641a7–18; *GA* II 1, 734a30–3; *Metaph.* I 8, 988b26–9, 989a25–6; XII 6, 1071b28–37.

24 On art, or skill, as an efficient cause: *Ph.* II 3, 195a5, b21–5; *GC* II 9, 335b32–5; *de An.* I 3, 407b26; *GA* 730b15–19; *Metaph.* VII 7, 1032b21–3; VIII 4, 1044a30–1; XII 4, 1070b28–9.

25 *Cael.* IV 3, 310a27–32; *GA* II 6, 743a18–27.

26 *Ph.* II 3, 195b16–21; *Metaph.* IX 8, 1049b24–7; XII 6, 1071b12–17. For an extensive discussion of Aristotle's causal powers, see S. Makin, "Commentary," in his edition of Aristotle, *Metaphysics Theta*, 17–269 (Oxford: Clarendon, 2006).

27 At *APo.* I 9, 76a24; 13, 78b37; and *Metaph.* XIII 3, 1078a16, Aristotle speaks of a "mechanical science" (μηχανική), which is subordinate to mathematics in the same way that astronomy, optics, and harmonics are. Sambursky, *Physical World of the Greeks*, 92; M. Jammer, *Concepts of Force: A Study in the Foundations of Dynamics* (Cambridge, MA: Harvard University Press, 1957), 37; and P. Distelzweig, "The Intersection of the Mathematical and Natural Sciences: The Subordinate Sciences in Aristotle," *Apeiron* 46 (2013): 85–105, agree with this claim inasmuch as they hold that many of the basic causal powers in Aristotle's mechanics are quantifiable.

28 This account of Aristotle's mechanics does not depend upon accepting Aristotle as the author of *Mechanica* (*Mechanical Problems*), which was attributed to him in antiquity. J. Anders, "*Problēmata Mēchanika*, the Analytics, and Projectile Motion," *Apeiron* 46 (2013): 106–35, argues persuasively that even if the *Mechanica* is not by Aristotle, it is Aristotelian, because it follows the method described by Aristotle in his *Organon* for dealing with such questions, and its discussion of projectile motion takes up where his *Physics* leaves off.

29 Sambursky, *Physical World of the Greeks*, 55, is one of the many commentators who claim that Aristotle has these two completely separate accounts of the physical world.

30 *Ph.* III 2, 202a3–9; VII 1, 242b59–63; 2, 243a32–5, 243a11–244b2; VIII 5, 256b18–20, 258a20–1; *GA* II 1, 734a3–5; 4, 740b22–741a4.

31 *Ph.* VII 2, 243a11–244b2.

32 *GC* I 10, 328a33–4. Williams, *Aristotle's De generatione et corruptione*, 150, comments on this passage: "Substances divided into small quantities mix better because they provide maximum surface for contact, and contact is necessary for the mutual action/passion which brings about the mixing." See also *Mete.* II 2, 355b25–33.

33 *Ph.* VIII 10, 267b7–8; *Mete.* II 4, 361a35–6.

34 *Mete.* I 4, 342a22–8. E. Hussey, "Aristotle's Mathematical Physics: A Reconstruction," in *Aristotle's* Physics*: A Collection of Essays*, ed. L. Judson, 213–42 (New York: Clarendon, 1995), esp. 221, argues that in this and other passages Aristotle uses parallelograms in combining different motions; that is, the actual motion of a body is represented as the result of the combination of two other motions, where the latter are represented as two

adjacent sides of a parallelogram. Since the two motions are caused by the exercise of two different powers, the parallelogram also shows how causal powers can be combined to produce a single motion. S. Berryman, *The Mechanical Hypothesis in Ancient Greek Natural Philosophy* (Cambridge: Cambridge University Press, 2009), 100, argues that the evidence is insufficient to say that Aristotle explicitly uses parallelograms.

35 *Ph.* IV 8, 215a14–19; VIII 10, 267a2–11; *Cael.* III 2, 301b18–30.

36 *Ph.* VIII 1, 250a17–251b10; 2, 253a7–20; 5, 256b3–13; 7, 260a26–b15.

37 J. Jammer, *Concepts of Mass in Classical and Modern Physics* (Cambridge, MA: Harvard University Press, 1961), 25, also attributes the principle of the conservation of matter to Aristotle.

38 *Ph.* VII 5, 249b27–250a28; VIII 10, 266a26–8; *Cael.* I 7, 275a2–10; III 2, 301b4–6; *GA* II 1, 732a19–20.

39 The most extensive modern treatment of these proportions is found in I. E. Drabkin, "Notes on the Laws of Motion in Aristotle," *American Journal of Philology* 59 (1938): 60–84. See also I.B. Cohen, *The Birth of a New Physics*, 2nd ed. (New York: Norton, 1985), 15–22; G.E.L. Owen, "Aristotelian Mechanics," in *Aristotle on Nature and Living Things*, ed. A. Gotthelf, 227–45 (Pittsburgh: Mathesis Publications, 1985); Hussey, "Aristotle's Mathematical Physics," 213–42; Berryman, *Mechanical Hypothesis in Ancient Greek Natural Philosophy*, 97–103.

40 Cohen, *Birth of a New Physics*, 19, captures this relation with the formula V (velocity) is proportional to F (force) divided by R (resistance); Hussey, "Aristotle's Mathematical Physics," 215, sums up these several proportions as an equation between the power of the agent times the temporal length of the change, on the one hand, and a constant times the amount of change and the size of the changing thing (adjusted for relative density), on the other.

41 *Ph.* VII 5, 250a12–19.

42 *Ph.* IV 8, 215a24–216a11.

43 Sorabji, *Matter, Space, and Motion*, 145n22, also reads these texts as presenting a reductio ad absurdum of the atomists' position and not as asserting that motion in a void would be instantaneous.

44 *Cael.* I 6, 273b28–274a18; III 2, 301a21–b18.

45 *Mete.* IV 6, 383a7–9.

46 *Mete.* IV 7, 384b3–4.

47 *GC* I 7, 324a9–14; *Ph.* VIII 5, 257b7–14; *Metaph.* II 1, 993b24–6; VII 9, 1034a21–5. Hankinson, *Cause and Explanation in Ancient Greek Thought*, 89–97 and 129, calls this the Principle of Causal Synonymy, attributing the term to Aristotle inasmuch as Aristotle sometimes says that the cause

and effect are synonymous; Hankinson also argues that this principle can already be found in Plato's *Phaedo* 100e–101b.

48 *GC* I 7, 324a5–14; *de An.* II 5, 417a19–21.

49 *Ph.* II 2, 194b13; 198a24–7; III 2, 202a9–12; *GC* I 5, 320b17–21; *Metaph.* VII 7, 1032a24–5; 8, 1033b32; IX 7, 1049a23–6.

50 *GC* I 5, 320b17–21; 7, 324a9–11; *Metaph.* II 1, 993b24–6; VII 7, 1032a24–5; 8, 1033b32; IX 7, 1049a23–6.

51 Owen, "Aristotelian Mechanics," esp. 239, argues that Aristotle does not have a general set of laws that apply to all physical bodies, because his proportionalities between force and resistance apply only to violent motion. For, Owen argues, Aristotle holds that natural motion always involves acceleration, whereas violent motions are treated as constant. W.D. Ross, *Aristotle's Physics* (Oxford: Oxford University Press, 1936), 26 and 30, makes a similar argument. Aristotle, however, does apply his proportionalities between force and resistance to both natural and violent motions, because it does not matter whether the causal powers producing these motions are internal or external to the bodies being moved. Moreover, acceleration and constant motion are found in both natural and violent motions, and Aristotle treats them in the same way, as arising from the proportion between the causal power applied and the resistance arising from the body and the medium through which it is passing. Thus, in the case of both acceleration and constant motion, the distinction between natural and violent motion is irrelevant to these proportionalities; Aristotle is interested here only in the ratio of causal power to resistance.

52 Sambursky, *Physical World of the Greeks*, xii; T. Kuhn, "Mathematical versus Experimental Traditions in the Development of Physical Science," in *The Essential Tension*, 31–65 (Chicago: University of Chicago Press 1977), esp. 55.

53 On this topic, see section 5.3 below.

54 *Cael.* II 7, 289a20–35; *Mete.* I 3, 341a13–32.

55 *Mete.* I 3, 341a19–29.

56 M. Leunissen, *Explanation and Teleology in Aristotle's Science of Nature* (Cambridge: Cambridge University Press, 2010), 168–71, argues that Aristotle also applies the same teleological principles to both celestial and terrestrial bodies. By way of example, she points to his argument that the stars lack feet or legs because they are fixed in the heavenly spheres and do not propel themselves forward; because nature does nothing in vain – a teleological principle – they lack such superfluous body parts. In this case, his claim about the stars is based upon his observation of self-propulsion in terrestrial animals.

57 *Mete.* II 3, 358b14–359a14. Similarly, he argues at *Ph.* IV 6, 213a25–32 that air is a real physical body, as opposed to just a void, because it can be used to inflate wineskins or be trapped in a water siphon between bodies of water.

Chapter Three

1 *Ph.* I 9, 192a31; II 3, 194b23–195a26; *GC* I 4, 320a2–3; *Metaph.* VII 7, 1032a17. Strictly speaking, he says it is "that out of which something comes to be, being present in it." This book argues that for Aristotle the material causes of perceptible objects are always matter of some kind, in the sense of physical stuff. In order to keep this claim from being merely tautological – the matter of perceptible objects is matter – I have translated *hyle* (ὕλη) as "material cause," and not as "matter," unlike many translations and much commentary on Aristotle. The translation of *hyle* as "matter" is particularly unfortunate, because material causes for Aristotle often have nothing to do with physical matter, or if they do, they are much more than just physical matter. This question is discussed further in my "Matter and Aristotle's Material Cause," *Canadian Journal of Philosophy* 31 (2001): 85–112.
2 *Ph.* II 3, 195a16–21.
3 *Metaph.* VII 10, 1035b31–1036a12; 11, 1036b35–1037a5. The intelligible material cause discussed in these passages individuates non-perceptible objects that are the same in kind, say, several mathematical circles; at *Metaph.* VIII 6, 1045a33–6, Aristotle seems to extend the notion of an intelligible material cause to the generic part of any definition, which takes us even farther from a physical material cause.
4 McMullin, "Introduction," 5–8, and "Matter as a Principle," 169–208, esp. 201–3, in *Concept of Matter*; Wieland, *Die aristotelische Physik* (1970), 140n29, 173–87, and 209–11; Kung, "Can Substance Be Predicated of Matter?" esp. 145–6; Frede, "On Aristotle's Conception of the Soul," 100; Code and Moravcsik, "Explaining Various Forms of Living," 139 and 143; Hankinson, *Cause and Explanation in Ancient Greek Thought*, 130.
5 In his *Substance, Form and Psyche*, 165, M. Furth defines a material cause as "something susceptible of those more specific determinations not excluded by the degree of differentiation it already incorporates." This definition is too broad, inasmuch as any kind of subject capable of further determination satisfies it – say, an animal being male or female – but it does capture nicely the feature of material causes that their susceptibility for further determination depends upon the properties they already have.

6 *Ph.* II 9, 200a7–13; *PA* I 1, 642a9–14; *Metaph.* VIII 4, 1044a27–9. This requirement holds not just for artefacts; at *De anima* I 1, 403b17–18, Aristotle says that the passions of the soul are inseparable from a "natural material cause" (φυσικὴ ὕλη).

7 *Metaph.* V 4, 1015a7–10; VIII 4, 1044a15–25; IX 7, 1049a18–27; *Ph.* II 1, 193a9–21.

8 This point is emphasized by P. Bogaard, "Heaps or Wholes: Aristotle's Explanation of Compound Bodies," *Isis* 70 (1979): 11–29; and Hankinson, *Cause and Explanation in Ancient Greek Thought*, 130.

9 *PA* II 1, 646a13–24; another example, at *Metaph.* XII 3, 1070a19, is fire, flesh, and the head.

10 *PA* II 1, 646a13–24; *GA* I 1, 715a5–12; *Mete.* IV 12, 389b26–8.

11 *Cael.* III 3, 302a16–28; *GC* I 5, 320b12–14; II 1, 329a8–13, 24–32.

12 In several places, Aristotle speaks of the "first material cause" (πρώτη ὕλη) of various perceptible objects (*Metaph.* V 4, 1015a7–10; VIII 4, 1044a15–25; IX 7, 1049a18–27; see also *Ph.* II 1, 193a9–21). Sometimes he uses this term to refer to the proximate material cause of something, what it is immediately made of, but he also uses this term for the ultimate material cause, the material cause that is first because it does not have a material cause of its own. Traditionally, commentators have used "prime matter" to mean the latter, namely the ultimate material cause of perishable objects.

13 *Ph.* III 5, 204b29–35; 5, 205a10–12, b25–31; IV 4, 211a2–6; 9, 217a21–6; *GC* I 5, 320b12–14; II 1, 329a8–13, 24–32; 5, 332a20–7, b1; *Cael.* III 6, 305a22–32; IV 5, 312b20–33.

14 *Metaph.* II 3, 995a14–18; VI 1, 1025b28–1026a10; *Ph.* II 2, 193b22–194a27; III 4, 203b30–204a2; *Cael.* III 1, 299a14–17; *de An.* I 1, 403a24–b19; III 7, 431b12–16; *PA* I 1, 641a25–7; *APo.* I 13, 79a7–10.

15 Their material cause is natural (φυσικὴ ὕλη): *De anima* I 1, 403b17–18; corporeal (ὕλη σωματική): *Metaph.* I 8, 988b23; movable in place (ὕλη τοπική): *Metaph.* VIII 1, 1042b6; 4, 1044b7–8; XII 2, 1069b24–6; and changeable (ὕλη κινητή): *Metaph.* VII 10, 1036a10–11. Imperishable perceptible objects have a material cause subject only to change of place, but not other changes; still, because locomotion is the fundamental kind of change (see section 1.1.3), every perceptible object must have at least this kind of material cause.

16 *Metaph.* II 3, 995a14–18.

17 *Ph.* II 2, 193b22–194a17.

18 Contrary to Mansion, *Introduction*, 154–5, and those who claim that the concept of matter, both now and in the past, is indistinguishable from that of a material cause, e.g., McMullin, *Concept of Matter*, 1–6.

19 *Cael.* I 1, 268a1–7; 9, 279a9; III 1, 298a27–b6.

20 *Ph.* II 2, 193b31–194a 7; 7, 198a17–18; *Cael.* III 1, 299a14–18; *de An.* I 1, 403b14–15; *Metaph.* I 8, 989b32–3; II 3, 995a14–18; VI 1, 1025b30–1026a10; VII 7, 1032a20–2; 10, 1035a24–b3; VIII 5, 1044b27–9.

21 *Ph.* II 1, 193a28–b8; 2, 194a12–27; 7, 198a22–4; 8, 199a30–1; *de An.* I 1, 403a24–b19; III 4, 429b13–14; *PA* I 1, 641a25–7; *Metaph.* V 4, 1015a6–11; VI 1, 1025b30–1026a6; VII 11, 1037a15–17.

22 *GC* I 7, 323b18–324a9; *Ph.* I 5, 188a30–4.

23 *GC* I 6, 322b11–21; *Ph.* I 7, 190b33–5.

24 *GC* I 7, 324a34–b7; 10, 328a19–22; *Ph.* I 7, 191a3–14.

25 *GC* I 7, 324b18–22.

26 *GC* I 7, 324b4–6.

27 *GC* I 7, 324a26–b4.

28 As we saw in the introduction, the traditional view of Aristotle's material cause holds that all of the permanent properties of perceptible objects are due to their formal cause, not their material cause: Owens, "Matter and Predication," 99–115; Charlton, *Aristotle's Physics I & II*, 75–7; Kosman, "Animals and Other Beings in Aristotle"; Gill, *Aristotle on Substance*, 146–67; Cohen, "Aristotle on Heat, Cold, and Teleological Explanation"; Katayama, *Aristotle on Artifacts*; Frey, "Organic Unity and the Matter of Man"; Scharle, "Material and Efficient Causes in Aristotle's Natural Teleology."

29 *GC* I 6, 323a9–12, 22–5.

30 *Cael.* I 9, 278a10–12, b3–4; at *Metaph.* VIII 1, 1042a25–b3, Aristotle says that all perceptible things have a material cause capable of underlying change.

31 *Ph.* IV 1, 208b22–5; *Metaph.* XIV 5, 1092a19–20.

32 *Ph.* IV 4, 212a6–7; 5, 212b28–9.

33 *Ph.* IV 2, 209b6–11 and 30–2.

34 *Cael.* III 8, 306b3–29; *GC* I 5, 320b14–17.

35 *Ph.* III 7, 207b34–208a4.

36 *Ph.* I 7, 190a13–191a5; *Metaph.* VII 8, 1033a24–31; 9, 1034b7–16.

37 *Ph.* I 9, 192a25–34; *Metaph.* VII 8, 1033a32–b5.

38 *Ph.* I 8, 191b13–15; 9, 192a3–6, 20–5.

39 *Ph.* I 7, 190b1–5; *Metaph.* VII 7, 1032b31–1033a1. Some commentators argue that Aristotle does not require a persisting substratum for generation and destruction, or what amounts to the same thing, that the transformation that takes place in these changes is so complete that nothing from the original object remains as a constituent part of what results: Aquinas, *Super libros De generatione et corruptione*, bk I, lect. 10, #80; McMullin, "Introduction," 7; FitzGerald, "'Matter' in Nature and the Knowledge of Nature," esp. 86; Cohen, "Aristotle on Heat, Cold, and Teleological

Explanation"; Gill, *Aristotle on Substance*, 146; M. Johnson, *Aristotle on Teleology* (Oxford: Oxford University Press, 2005), 144; D. Henry, "Substantial Generation in Physics I 5–7," in *Aristotle's* Physics*: A Critical Guide*, ed. M. Leunissen, 144–61 (Cambridge: Cambridge University Press, 2015), esp. 155. Chapter 1 above argues that this view is incompatible with Aristotle's claims that change is always more than just temporal succession and never simply the replacement of one object by a brand new one. Following the account of Aristotle's mechanics in chapter 2 above, this chapter argues that his account of the dynamics of generation and destruction – how efficient causes bring about generation and destruction – also requires a persisting, physical substratum for these changes.

40 *Ph.* I 9, 192a25–34. In this passage, Aristotle states that the substratum is present "not accidentally" in the thing generated; his point here is that whatever is identified as the substratum of a change should be the proximate substratum of that change, and not something only accidentally related to it. This is compatible with my claim that the formal cause of a generated object is accidental to the substratum of generation and destruction in the sense of not being an essential part of whatever acts as that substratum.

41 *Ph.* I 9, 192a31–2; *GC* I 4, 320a2–3; *Metaph.* VII 7, 1032a15–20, 1033a8–10; 8, 1033b16–19; 10, 1035a25–30; 15, 1039b29–30; VIII 4, 1044b8–11; 5, 1044b27–9.

42 *Metaph.* VII 7, 1032b30–3a5; 8, 1033a24–b19; 9, 1034b7–19; VIII 1, 1042a29–31; XII 3, 1069b35–70a4.

43 *GA* I 22, 730a33–b19. This claim is independent of the dispute about whether the male or the female supplies the material cause in the case of sexual generation. Lennox, "Teleology, Chance, and Aristotle's Theory of Spontaneous Generation," *Journal of the History of Philosophy* 20 (1982): 219–38, argues that the latter is Aristotle's position; J. Rist, *The Mind of Aristotle* (Toronto: University of Toronto Press, 1989), 223–6, argues that Aristotle changed his mind from the former to the latter.

44 *Metaph.* VII 8, 1033a24–b19; IX 8, 1049b17–29; *GA* I 19, 727b31–3; 20, 729a11–32; 21, 730a14–27; II 1, 734b20–3; 4, 740b24. This is the view of Mansion, "The Ontological Composition of Sensible Substances." Lennox, "Teleology, Chance, and Aristotle's Theory of Spontaneous Generation," 223, calls natural generation the "replication" of a formal cause in a material cause.

45 *Ph.* I 7, 190b1–17, 23–191a3; 9, 192a3–6, 16–25; *Metaph.* V 4, 1014b26–35. At *Ph.* I 9, 192a32; II 3, 194b24; *GA* I 18, 724a25–7; and *Metaph.* VII 7, 1032b32, Aristotle says that the material cause "exists in" (ἐνυπάρχειν) the generated substance. At *GA* I 18, 724a23–7, he says the material cause is "shaped" (σχηματισθέντος) during the process of generation.

46 *Ph.* I 9, 192a3–6, 20–5.
47 *Ph.* I 5, 188a30–4; VII 4, 249a2–3; *GA* II 6, 743a21–7; *Metaph.* VIII 4, 1044b1–3; IX 7, 1049a1–3, 14–18.
48 *de An.* II 1, 412a7–9; *Metaph.* VII 7, 1032a20–2; 15, 1039b27–31; VIII 1, 1042a27–9; IX 7, 1049a23; 8, 1050a15, b27; XII 2, 1069b14; 4, 1070b12–13; 5, 1071a10; 10, 1075b22. At *Metaph.* VI 1, 1026a2–4; VII 7, 1032a20–2; VIII 5, 1044b27–9; XII 2, 1069b24–6, he says that everything subject to change has a material cause (ὕλη).

Chapter Four

1 *Metaph.* V 4, 1015a7–10; VIII 4, 1044a15–25; IX 7, 1049a18–27; *Ph.* II 1, 193a9–21.
2 In the first instance, "prime matter" refers to the ultimate material cause of perishable objects. Given the layering of material causes, however, this ultimate material cause also serves as the proximate material cause of the material elements.
3 *Cael.* I 2, 268b27–269a2, 269a30–b6; III 2, 300a20–b8; *GC* II 2, 329b24–34; *Mete.* IV 1, 378b10–14; *PA* II 2, 648b9–11.
4 *GC* II 2, 329b7–24.
5 *GC* II 3, 330a30–b5; *Mete.* IV 1, 378b10–14; *PA* II 2, 648b9–11.
6 *GC* II 2, 329b24–32; *Mete.* IV 1, 378b10–14.
7 *GC* I 6, 323a9–12, 22–5; 7, 323b29–34; II 2, 329b7–10; *Ph.* IV 1, 209a17–18.
8 *GC* I 3, 319a29–b4; 7, 323b29–33; *Ph.* I 7, 190b33.
9 *GC* I 7, 324b4–9; II 1, 329a8–11, 24–35; *Ph.* IV 9, 217a21–6; 1, 209a17–18. D. Graham, "The Paradox of Prime Matter," *Journal of the History of Philosophy* 25 (1987): 475–90, esp. 487, argues that the sublunary elements have no physical attributes in common. One obvious counter-example is extension. This, he says, belongs to the elements by virtue of their respective formal causes, not their common material cause; the latter, prime matter, cannot be defined in terms of extension, because it lacks any more basic attributes by virtue of which it would be extended; for example, the modern notion of mass. This objection misses the mark; even if extension is a derivative attribute, the question is whether it is necessary to prime matter. If it is, then it cannot be separated from prime matter and still leave anything behind: either prime matter exists as extended or it does not exist at all. If extension is not necessary to prime matter, it must be possible to add the defining properties of the elements – heat, cold, fluidity, and solidity – to an unextended substratum. This, Aristotle holds, is impossible.
10 *Ph.* III 1, 200b12–25; *Metaph.* XIII 3, 1077b22–7.

11 *Ph.* VI 4, 234b10–235a10, b1–5; VII 1, 242a40; VIII 6, 258b24–5; 10, 267a22–3; *Cael.* I 1, 268a7–11.
12 *Ph.* IV 4, 212a6–7; 5, 212b28–9.
13 *Ph.* IV 1, 209a6–7; 6, 213b20; *GC* I 5, 321a7–9, b15–16; *de An.* II 7, 418b17.
14 *GC* I 2, 315b24–33; 5, 320b14–17, 22–28; II 1, 329a13–24; *Cael.* III 1, 299a2–12; 7, 306a23–30.
15 *Ph.* VIII 7, 260b11–13, 261a1–7; *GC* II 10, 336a20–6; *Metaph.* VIII 1, 1042b2–6.
16 *Metaph.* VIII 1, 1042b6.
17 *Ph.* IV 2, 209b6–11.
18 For the traditional view of prime matter as pure potentiality, see the introduction.
19 Several commentators have tried to dispense with a purely potential prime matter by arguing that it is the sublunary elements themselves, or these elements with just one of their two defining characteristics, that act as the ultimate material causes of perishable perceptible substances: H.R. King, "Aristotle without *Materia Prima*," *Journal of the History of Ideas* 17 (1956): 370–89; B. Jones, "Aristotle's Introduction of Matter," *Philosophical Review* 83 (1974): 474–500; Charlton, *Aristotle's Physics I & II*, 129–45; and Charlton, "Prime Matter: A Rejoinder," *Phronesis* 28 (1983): 197–211; Furth, *Substance, Form and Psyche*, 221–7; Gill, *Aristotle on Substance*, 42–6 and 65–7. See also N. Lobkowicz, "Discussion," in McMullin, *Concept of Matter*, 120 and A. Wolter, "The Ockhamist Critique," in McMullin, *Concept of Matter*, 144–6. In effect, these commentators deny that there is a material substratum prior to the material elements. For the reasons set out above, however, the traditional doctrine is correct that the generation of the elements requires a substratum more primitive than the material elements.
20 Thus, contrary to King, "Aristotle without *Materia Prima*"; Charlton, *Aristotle's Physics I & II*, 129–45; Charlton, "Prime Matter"; and Gill, *Aristotle on Substance*, one of the four contraries does not need to persist whenever one sublunary element is generated from another. Gill argues that the material elements are not composite substances, with a material and formal cause of their own, and that the role of the substratum persisting through the generation and destruction of these elements is taken over by their defining contraries because the nature of a composite substance is "exhausted by its form" (40, 65–7, 163–7). According to Aristotle, however, change and generation require a persisting substratum that is extended, movable, and corporeal in its own right, independent of the four defining contraries of the sublunary elements. Thus, as some commentators have argued, prime matter must be extended and occupy space: R. Sokolowski, "Matter, Elements and Substance in Aristotle,"

Journal of the History of Philosophy 8 (1970): 263–88; J.W. Dye, "Aristotle's Matter as a Sensible Principle," *International Studies in Philosophy* 10 (1978): 59–84; S. Cohen, "Aristotle's Doctrine of the Material Substrate," *Philosophical Review* 93 (1984): 171–94. According to Sorabji, *Matter, Space, and Motion*, 14n34, the view that prime matter is extended is at least as old as Simplicius, the sixth-century CE commentator on Aristotle.

21 *Ph.* III 5, 204b29–35; 5, 205a10–12, b25–31; IV 4, 211a2–6; 9, 217a21–6; *Cael.* III 6, 305a22–32; IV 5, 312b20–33; *GC* I 5, 320b12–14; II 1, 329a8–13, 24–32; 5, 332a20–7, b1. At *De caelo* IV 5, 312b20–33, Aristotle argues that there cannot be just one kind of matter, because there must be as many different kinds of matter as there are natural motions, and we observe several different natural motions in material substances. This is an argument against the view that there is a single, separately existing element out of which every other perceptible object is made; it does not preclude the existence of a common physical stuff that never exists apart from the elements. Contrary to the claim made by Graham, "Paradox of Prime Matter," 487, this passage does not imply that any physical substratum common to the elements would have to be itself already heavy or light; heaviness and lightness arise only once the contrary, differentiating properties of the elements have been added to their common substratum.

22 *GC* I 4, 319b6–24.

23 While agreeing that the generation of the sublunary elements requires a persisting substratum that is extended and movable, Cohen, "Aristotle's Doctrine of the Material Substrate," argues that Aristotle does not require a persisting substratum for every change. His principal reason is that the proximate material cause of animate substances, typically their organs, does not persist through their generation and destruction. Nevertheless, if it is the case that locomotion requires a persisting substratum and every other kind of change presupposes locomotion in the thing undergoing that change, then every change will require a persisting substratum. This persisting substratum may be at a lower level than the proximate material cause of the substance in question, but, whatever else that substratum may be, it must be extended, corporeal matter. Corporeal matter may be only the proximate material cause of the sublunary elements, but it must also be a part of the substratum of every other change. This does not deny Cohen's claim that Aristotle is typically interested in the proximate material cause when discussing the generation of substances, and that this is usually something quite specific and more complex than the sublunary elements. Still, physical matter and locomotion retain a fundamental role in the explanation of all change, not just the behaviour of the sublunary

elements, and this fundamental role is sufficient to guarantee that every change has a persisting, physical substratum.

24 Gill, *Aristotle on Substance*, 42–6, criticizes the view that prime matter is pure potentiality, because such a substratum cannot ensure continuity in essential change. On the other hand, she argues that if the persisting substratum has any essential properties of its own, then the change in question is no longer essential, but accidental. Her solution to this difficulty is to say that the substratum of generation persists merely potentially in the resulting object. A substratum that persists merely potentially, however, is insufficient. In fact, according to her own analysis of the generation of the elements, it is not prime matter that acts as the persisting substratum, but one of the defining powers of the elements, which is somehow transferred to a new subject (146–60). As we saw in chapter 3, however, this is impossible, because these powers cannot exist apart from an actual physical substratum, the continued existence of which is necessary for the persistence of these powers.

25 *Ph.* I 3, 186a28–34; 7, 190a14–17, b23–5, b36–191a3; 9, 192a3–6, 22–5; *GC* I 3, 319a29–b4. At *Ph.* I 7, 190b11, Aristotle makes the related point that everything changeable is composite.

26 *Ph.* I 9, 192a1–25.

27 For a discussion of the arguments in favour of a purely potential prime matter drawn from Aristotle's metaphysical works, see my "Prime Matter and Actuality," *Journal of the History of Philosophy* 33 (1995): 197–224.

Chapter Five

1 *Metaph.* V 5, 1015a34–6, b11–15; *APo.* I 6, 74b5–7; 33, 88b32–3; *GC* II 11, 337b35–338a2; *EN* VI 3, 1139b19–24. These passages also suggest that Aristotle grounds eternity in necessity: something is eternal just in case it cannot be otherwise.

2 *Metaph.* V 5, 1015a20–6, b3–9; *Ph.* II 9, 199b34–5, 200a5–30; *GC* II 11, 337b14–33, 338b6–11; *PA* I 1, 639b21–640a1. At *Metaph.* V 5, 1015a26–8; and XII 7, 1072b7–13, he distinguishes a third kind of necessity, what happens by force or external compulsion, also known as violent motion or change. Since this kind of necessity deals with motion contrary to the natural motion of a body, it is not so much a different kind of necessity, as a particular kind of motion, one caused in a certain way. On this topic, see section 6.2 below.

3 *Ph.* II 9, 200a7–15; *PA* I 1, 642a2–14; *Metaph.* VIII 4, 1044a27–9.

4 *PA* I 1, 640a3–8; *GC* II 11, 337b14–33, 338b5–11.

5 *Cael.* I 2, 268b15–24; 9, 278b22–279a5; III 3, 302b5–9; *Ph.* IV 1, 208b8–21.

6 *APo.* II 12, 96a2–5; *Ph.* II 9, 199b34–200a5, 30–2; VIII 1, 252a17–19; *Cael.* IV 2, 308b12–15; 4, 311b14–19; *GC* II 11, 338a17–b11; *PA* I 1, 642a33–b4; *EN* II 1, 1103a18–23.

7 *Ph.* III 5, 205b31–5; IV 4, 211a3–6.

8 *Ph.* VIII 6, 258b10–259a13; *Metaph.* XII 7, 1072a19–b13; 10, 1075a11–15.

9 Aristotle sometimes *contrasts* what happens as the result of simple physical necessity with what happens for the sake of an end or good, e.g., *GA* I 4, 717a15–17; II 1, 731b20–4; V 1, 778a29–b2, 12–19; 8, 789b19–22; *PA* I 1, 639b21–640a9; III 2, 663b13–15; *GC* II 11, 337b14–35. As the natural motions of the elements show, however, these two types of change are not mutually exclusive. Not every case of teleological change is a case of conditional necessity.

10 *Ph.* IV 8, 214b12–17, b28–215a14; *Cael.* I 9, 278b9–279a12; III 2, 300b17–26.

11 D.M. Balme, "Greek Science and Mechanism: 1. Aristotle on Nature and Chance," *Classical Quarterly* 33 (1939): 129–38: and Balme, *Aristotle's De partibus animalium I and De generatione animalium I* (Oxford: Clarendon, 1972), 76–84; Ross, *Aristotle's Physics*, 43; R. Demos, "The Structure of Substance According to Aristotle," *Philosophy & Phenomenological Research* 5 (1944–5): 255–68; J. Owens, "The Teleology of Nature in Aristotle," *Monist* 52 (1968): 159–73, reprinted in *Aristotle: The Collected Papers of Joseph Owens*, ed. J. Catan, 136–47 (Albany, NY: SUNY Press, 1981); Wieland, *Die aristotelische Physik*, 265. R. Sorabji, *Necessity, Cause and Blame* (London: Duckworth, 1980), 144n3, lists other commentators, both ancient and modern, who argue that Aristotle allows for only hypothetical necessity in nature. Against this view, I. Bodnar, "Movers and Elemental Motions in Aristotle," *Oxford Studies in Ancient Philosophy* 15 (1997): 81–117; and M. Scharle, "Elemental Teleology in Aristotle's *Physics* 2.8," *Oxford Studies in Ancient Philosophy* 34 (2008): 147–83, have argued that the natural motions of the material elements are both unconditionally necessary because they follow an eternal pattern and take place for the sake of an end independent of the ends of the objects made from them.

12 *Ph.* IV 1, 208b8–21; 4, 211a2–6; *Cael.* I 2, 268b15–269a2; III 3, 302b5–9.

13 *Cael.* IV 1, 308a14–30; 2, 308b12–15.

14 *Cael.* II 14, 296b7–18.

15 *Ph.* III 5, 205b30–5; *Cael.* IV 3, 310a32–b10.

16 At *PA* I 1, 640a6–8, Aristotle says that it is not possible to trace the hypothetical necessity involved in processes of generation back to anything eternal.

17 *Cael.* IV 1, 307b32–3; 308a14–30; 2, 308b12–15; 4, 311a16–29, b14–19; *Ph.* VIII 4, 255b11–12.

18 *Cael.* IV 2, 308b18–28.
19 *Cael.* IV 2, 309b13–16; 4, 311a20–3.
20 *Metaph.* V 5, 1015a26–8; XII 7, 1072b7–13; *Cael.* III 2, 301b17–30.
21 *Mete.* I 4, 341b1–342a34; 9, 346b16–36, 347a8–12; 10, 347a13–28; 12, 348b2–31; III 1, 370b3–371a18.
22 On this topic, see chapter 6 below.
23 *Ph.* IV 4, 211a2–6; *Cael.* IV 3, 311a8; 4, 311b14–19.
24 On this topic, see section 5.3 below.
25 *EN* II 1, 1103a18–23.
26 *GC* II 11, 338a17–b11.
27 *Cael.* I 2, 269a2–18.
28 *GC* II 2, 329b7–24; I 6, 322b21–9; 323a10–12.
29 *GC* II 2, 329b18–24.
30 *GC* II 2, 329b32–330a29.
31 *GC* II 2, 329b24–32; 3, 330a30–331a6; *Mete.* IV 1, 378b10–14; *PA* II 2, 648b9–11.
32 *GC* II 7, 334a26–b19.
33 *Ph.* II 2, 194a34–6; 3, 195a23–6; 7, 198b8–9; *GA* II 1, 731b23–4; V 8, 789b5–6; *Metaph.* V 2, 1013b25–8; XII 7, 1072b2–3.
34 *Ph.* II 2, 194a30–3.
35 *PA* II 2, 649a19–21.
36 *PA* I 1, 642a32–b2 describes that part of the process of respiration in which air moves in and out of the body through successive heating and cooling as taking place by necessity and not for the sake of an end; at IV 2, 677a1–18, he says the same about many of the residues produced in the bodies of animals. As we shall see in chapter 9, these changes can certainly be made to be useful, but only by bringing in other things and what benefits them.
37 At *Metaph.* I 3, 984b11–14, Aristotle says that it is not likely that fire or earth or any other such thing can be the cause of things being or coming to be good or noble. At *GA* V 1, 778a29–b2, and V 8, 789b19–22, he indicates that Book V of that work deals with those characteristics of animals that do not come to be for the sake of an end or from their formal cause, but as a result of the material causes of animals and the efficient causes found in those material causes.
38 *GA* II 1, 731b28–31; *GC* II 10, 336b27–9.
39 *Ph.* I 4, 188a15–17.
40 Johnson, *Aristotle on Teleology*, argues that these elemental, natural powers are always exercised for the immediate benefit of the bodies that have them. Apart from generation and destruction, however, it is hard to see

what those benefits might be. On Johnson's account, non-teleological causality does not operate in nature prior to, or alongside of, these end-directed causal powers; all causality involves the goal-directed exercise of natural powers. As a result, Aristotle's science of nature has no room for non-teleological causality of any kind.

41 *Ph.* VIII 4, 255a5–11.

42 *Mete.* IV 3, 380a23–6; *GC* II 5, 332a21–2.

43 *Ph.* IV 5, 212b2–3; 9, 217b11–12, 24–6; VIII 7, 260b7–10; *Cael.* III 1, 299b7–9. M.J. White, "Concepts of Space in Greek Thought," *Apeiron* 29 (1996): 183–98, argues that Aristotle has a relational account of natural place in that the natural place of an element is the one in which it is next to, or surrounded by, the right next element. In that case, he asks, why not an inverted world, with fire at the centre and earth at the periphery, next to ether (192). Part of the answer has to do with the connection between natural motion and relative density: because lighter elements are rarer and require more space for the same amount of matter, they will tend to move into places where they can occupy more space. In a spherical universe, that will mean moving away from its centre. At the same time, denser bodies, requiring less space, will be pushed toward the centre.

44 *Ph.* VIII 7, 260b7–10.

45 *Mete.* I 4, 341b11–13 and 342a16–17; *PA* II 7, 653a3–8.

46 *Cael.* II 7, 289a20–35; *Mete.* I 3, 341a13–32; 4, 341b18–24.

47 *Mete.* I 3, 341a19–29.

48 *de An.* II 7, 418b4–13.

49 *de An.* II 7, 418b11–13.

50 *de An.* II 7, 419a13–15; III 1, 424b27–30; 13, 435a12–19.

51 *Cael.* II 9, 290b30–291a28.

52 A. Falcon, *Aristotle and the Science of Nature* (Cambridge: Cambridge University Press, 2005), 11, argues that for Aristotle the direction of explanation is only from the celestial to the sublunary world; he is correct about the direction of causation, but not explanation: Aristotle explains several features of the celestial world by appealing to processes in terrestrial bodies.

Chapter Six

1 Aristotle makes this claim frequently: *Ph.* II 1, 193a28–31; 2, 194a12–27; 8, 199a30–2; *GC* I 5, 321b19–22; *de An.* II 2, 414a17; III 4, 429b13–14; *PA* I 1, 641a25–8; *GA* I 18, 724a20–b1; *Metaph.* V 4, 1015a6–11; VI 1, 1025b28–1026a6; VII 7, 1032a12–22; 8, 1033b12–26; 10, 1035b27–32; 11, 1037a5–7, 10–20.

2 See the notes to section 5.2 for references to commentators who argue there can be no simple physical necessity governing all perceptible objects as the result of the material elements. Balme, *Aristotle's De partibus animalium I*, 82, for example, argues this must be so because all properties of the elements and, indeed, their very existence, are necessitated by the final and formal causes of the natural substances made out of them, and whatever is necessitated hypothetically in this way can no longer be explained in terms of simple physical necessity. Kosman, "Animals and Other Beings in Aristotle," makes a similar argument about the parts of animals. Later, however, Balme came to the view that the interactions of the elements are governed by simple necessity and that these interactions explain at least some of the properties of the parts of animals: "Teleology and Necessity," in *Philosophical Issues in Aristotle's Biology*, ed. A. Gotthelf and J. Lennox, 275–85 (Cambridge: Cambridge University Press, 1987).

3 *Ph.* III 5, 205a10–12; IV 1, 208b8–21; 4, 211a2–6; *Cael.* I 2, 268b15–30; 9, 278b22–279a5.

4 *Cael.* I 2, 269a1–2; IV 4, 311a30–b8; *GC* II 8, 334b31–4.

5 *Ph.* VIII 4, 254b18–20; *Cael.* III 3, 302b5–9; *PA* II 1, 646a14–21.

6 *Cael.* I 2, 268b17–269a7.

7 *Cael.* I 2, 269a7–18; 3, 269b30–270a13.

8 *Cael.* II 1, 284a15–18.

9 Without limbs to raise them up, animals would naturally lie on the ground (*PA* IV 10, 686a27–687a2). Given the role of their roots, the natural place of plants is not just on the ground, but partially in the ground (*de An.* II 1, 412b2).

10 *Mete.* IV 4, 382a6–8.

11 Thus, the natural motions of the elements are explained in the same way wherever they occur, including inside animals: just as warm air in the atmosphere rises, cools, condenses into water, and falls to the Earth, so too vapours rise inside animals, are cooled and condense, and then flow downward (*PA* II 7, 652b34–653a8; *Somn.* 3, 457b30–458a6).

12 *Metaph.* VI 2, 1026b31–3.

13 *Cael.* I 12, 283b21–2; II 6, 288b15–19; *de An.* II 4, 416a6–8.

14 *Cael.* I 9, 279a7–12.

15 *Cael.* IV 6, 313b7–22.

16 *Cael.* II 6, 288b15–19; *de An.* II 4, 416a6–8.

17 *GC* II 8, 334b31–335a9; *Mete.* IV 4, 382a4–8; *GA* II 6, 743a3–20.

18 *Mete.* IV 4, 381b31–382a3.

19 *Mete.* IV 5, 382a22–7.

20 *GC* II 2, 329b32–4; *Mete.* IV 5, 382a22–7; 12, 390b6–10; *PA* II 1, 646a17–21.

21 *GC* II 2, 329b24–34; *Mete.* IV 1, 378b15–26; 4, 381b24–7; 5, 382a27–b6, 23–4; 8, 384b24–34, 385a23–8; 10, 388a23–5; 12, 390b4.
22 *Mete.* IV 4, 382a3–4; 5, 382b3–4; 10, 388a23–5; *PA* II 1, 646a14–21; II 2, 648b9–11.
23 *Mete.* IV 4, 382a3–8; 5, 382b11–23; 6, 382b32–383a16, 28–33; 7, 383b18–26; 10, 388a10–29; *PA* II 2, 649a30–4.
24 *Mete.* IV 5, 382b2–6; 6, 382b32–383a11; 7, 383b18–21; 8, 385a23–8; 10, 388a20–389a23; *PA* II 2, 649a30–2. These processes are discussed in greater detail in G. Freudenthal, *Aristotle's Theory of Material Substance: Heat and Pneuma, Form and Soul* (Oxford: Clarendon, 1995), 150–78.
25 *Mete.* IV 10, 389a7–9. Unlike other metals, iron is composed of more earth than water, and yet, like them, is softened by heat and hardened by cold. It is, however, only an apparent exception to his rule that earthy compounds are hardened by heat, because the softening that initially takes place when iron ore is heated is caused by the water and other impurities in it being driven out. Once it has been sufficiently heated, it becomes harder (*Mete.* IV 6, 383a27–b9). Here, as elsewhere, Aristotle sees water as crucial to the softening and solubility of physical objects; see also *Mete.* IV 6, 383a15–17; 7, 383b18–19, 384a3–12, 34–b23.
26 *Mete.* IV 7, 383b22–384a3; *GA* II 2, 735a30–736a24.
27 *Mete.* IV 7, 384a3–b24. At *PA* II 4, 650b14–651a12, he explicitly connects the congealing and liquefaction of the blood in animals to the amount of earth and water found in it.
28 *Mete.* IV 1, 378b15–20; 4, 382a4–8; 5, 382a22–b1; 8, 384b24–35; 10, 388a10–29.
29 *Mete.* IV 2, 379b18–380a7.
30 Given this mastering of the raw materials, some commentators have considered concoction to be analogous to a chemical reaction, e.g., H.H. Joachim, "Commentary," in Aristotle, *De generatione et corruptione* (Oxford: Clarendon, 1922), 175–7; I. Düring, *Aristotle's Chemical Treatise: Meteorologica* IV (Göteborg: Elanders, 1944), 10–12.
31 *Mete.* IV 2, 379b12–14; 3, 381b3–5.
32 *GA* II 6, 743a18–21.
33 *Mete.* IV 3, 381b6–8; *De juv.* 4, 469b12–14; *PA* II 3, 650a3–7, 14–31.
34 *GA* II 6, 743a3–20.
35 *de An.* II 4, 416b3–11; *GC* I 5, 321b35–322a16.
36 *GA* I 18, 725a3–21, b22; 19, 726b3–5.
37 *GA* I 19, 726b5–12 and 726b31–727a4, 727b31–3; 20, 729a28–33. From the blood, or its analogue in bloodless animals, the rest of a biological organism is generated: *PA* II 3, 650a34–b3; 4, 651a13–15; *GA* I 19, 726b3–12.

38 *de An.* II 4, 415a22–6, 416a19. Also the matter from which a biological organism is generated is the same in kind as the food from which it nourishes itself: *GA* II 4, 740b30–741a2.

39 *Mete.* IV 1, 378b28–379a1, 17–33; *Long.* 5, 466a18–b4; *PA* II 2, 648b2–11.

40 *PA* II 4, 650b19–651a19.

41 *PA* II 3, 650a33–b8.

42 C. Freeland argues that Aristotle also sees blood as the persisting matter for the generation of blooded animals, including humans: "Aristotle on Bodies, Matter, and Potentiality," in *Philosophical Issues in Aristotle's Biology*, ed. A. Gotthelf and J. Lennox, 392–407 (Cambridge: Cambridge University Press, 1987).

43 *PA* II 2, 648a2–11.

44 M. Leunissen, "Aristotle on Natural Character and Its Implications for Moral Development," *Journal of the History of Philosophy* 50 (2012): 507–30, esp. 513–20, considers in greater detail the connection between the material constituents of animals, including human beings, and their character traits, in particular the dependence of the latter on the former.

45 *PA* II 7, 652b34–653a8.

46 *GA* V 1, 779b13–35.

47 *de An.* II 7, 418b4–13.

48 *GA* V 7, 787b20–6, 788a3–7, 17–24.

49 *Cael.* II 9, 290b30–291a28.

50 *GA* V 1, 778a16–b2, 13–19; 780b1–13.

51 M. Leunissen and A. Gotthelf, "What's Teleology Got to Do with It: A Reinterpretation of Aristotle's *Generation of Animals V*," *Phronesis* 55 (2010), 325–56, agree that material necessity in the elements is the primary cause of the traits discussed by Aristotle in *GA* V, which are not the primary traits of animals and are subject only to what they call "secondary teleology," that is, the generation of secondary organs that perform functions non-essential but useful to animals, e.g., the production of teeth to enhance digestion and self-nourishment. The physical necessity at work in the elements, however, cannot be limited to the processes Leunissen and Gotthelf group under secondary teleology; the interactions of the elements cannot be necessary simply if they occur only sometimes and not others. Simple physical necessity must also be at work in primary teleology, the generation of the main organs and body parts without which a biological organism cannot live.

52 For further defence of this view, see R. Friedman, "Matter and Necessity in *Physics* B 9, 200a15–30," *Ancient Philosophy* 3 (1983): 8–11; M. Bradie and F. Miller Jr, "Teleology and Natural Necessity in Aristotle," *History of*

Philosophy Quarterly 1 (1984): 133–45; Balme, "Teleology and Necessity"; J. Lennox, "Material and Formal Natures in Aristotle's *De partibus animalium*," in *Aristotelische Biologie: Intentionen, Methoden, Ergebnisse*, ed. W. Kullmann and S. Föllinger, 163–81 (Stuttgart: Steiner 1997), reprinted in J. Lennox, *Aristotle's Philosophy of Biology: Studies in the Origins of Life Science* (Cambridge: Cambridge University Press, 2001), 182–204; C. Byrne, "Compositional & Functional Matter: Aristotle on the Material Cause of Biological Organisms," *Apeiron* 48 (2015): 387–406.

53 *GC* I 4, 320a2–3; *Ph*. I 9, 192a31–2; *Metaph*. VII 7, 1032a15–22, 1033a8–10; 8, 1033b16–19; 10, 1035a25–9; 15, 1039b29–30; VIII 4, 1044b8–11; 5, 1044b27–9.

54 *Metaph*. VII 7, 1032b30–3a5; 8, 1033a24–b19; 9, 1034b7–19; 10, 1035a25–9; XII 3, 1069b35–70a4.

55 See *Ph*. I 9, 192a32; *GA* I 18, 724a25–7; and *Metaph*. VII 7, 1032b32, where Aristotle says that the material cause "exists in" (ἐνυπάρχειν) the generated substance.

56 *GC* I 10, 327b22–31; II 7, 334a26–35; *Cael*. III 3, 302a15–25; see also *PA* II 1, 646a13–24.

57 Owens, "Matter and Predication"; Charlton, *Aristotle's Physics I & II*; Gerson, "Artifacts, Substances, and Essences"; Cohen, "Aristotle on Heat, Cold, and Teleological Explanation," esp. 259; and Gill, *Aristotle on Substance*, 146–64. Johnson, *Aristotle on Teleology*, 144, writes that the elements are "completely transformed" when they form homogeneous bodies such as bronze, flesh, or blood, and that this transformation happens again when homogeneous bodies are made into heterogeneous bodies such as faces and hands.

58 *GC* I 10, 327b5–6, 24–30; 328b18–22; II 7, 334b4–20. On the new sense of "potentiality" introduced by Aristotle in this context, see the notes on the above passages in Williams, *Aristotle's De generatione et corruptione*.

59 *GC* I 10, 328b22.

60 *GC* I 10, 328a11–16; II 7, 334a18–b2.

61 *GC* II 1, 329a24–b3; 3, 330a30–331a6; *Mete*. IV 1, 378b10–14; *PA* II 2, 648b9–11.

62 *GC* I 10, 328a28–33, b22; II 7, 334b8–19; *Cael*. III 3, 302a20–5.

63 *GC* I 10, 328a29.

64 Aristotle emphasizes that the elements are found potentially in homogeneous mixtures, but not the converse; for the elements can be separated out of these mixtures, but the latter cannot be separated out of the elements (*Cael*. III 3, 302a20–5; *GC* I 10, 327b27–31).

65 This point is nicely made by Bogaard, "Heaps or Wholes." See also K. Fine, "The Problem of Mixture," *Pacific Philosophical Quarterly* 76 (1995):

266–369; J. Bogen, "Fire in the Belly: Aristotelian Elements, Organisms, and Chemical Compounds," *Pacific Philosophical Quarterly* 76 (1995): 370–404; and A. Code, "Potentiality in Aristotle's Science and Metaphysics," *Pacific Philosophical Quarterly* 76 (1995): 405–18.

66 Aristotle's account of the non-homogeneous mixtures involved in meteorological events follows the same pattern: he explains shooting stars, rainfall, dew, hail, wind, thunderstorms, lightning, hurricanes, and typhoons by appealing to the basic properties of the elements out of which they are made. See *Mete.* I 4, 341b1–342a34; 9, 346b16–36, 347a8–12; 10, 347a13–28; 12, 348b2–31; III 1, 370b3–371a18. At *Mete.* IV 1, 378b15–17, he says that hot and cold are the basic active forces at work in all things, both homogeneous and non-homogeneous.

67 *Mete.* IV 1, 379a1–3; *GA* IV 4, 770b15–17. With respect to the authenticity of Book IV of the *Meteorologica*, H.D.P. Lee argues persuasively that, at most, chapters 8 and 9 are suspect: see his "Preface to Second Edition" and "Introduction," in his Loeb edition of Aristotle's *Meteorologica* xiii–xxi (Cambridge, MA: Loeb, 1962).

68 *Mete.* IV 2, 379b12–20; 379b33–380a9.

69 *Mete.* IV 1, 378b10–379a11; 4, 381b24–382b7; *PA* II 2, 648b2–8; *GA* II 6, 743a37–b1. These powers are also responsible for several natural processes that take place within already existing mixtures, e.g., ripening, boiling, baking, and digesting: *Mete.* IV 2–3, 379b10–381b22.

70 *Mete.* IV 1, 378b10–14; *PA* II 2, 648b9–11.

71 *Mete.* IV 2, 379b33–380a1; 12, 390b10–14; *GA* II 1, 734b27–8 and b36–735a5. This topic is extensively discussed in Freudenthal, *Aristotle's Theory of Material Substance*, 7–47.

72 *GC* I 10, 327b30–1; 327b5–6, 24–30; 328b18–22; II 7, 334b4–20.

Chapter Seven

1 *Ph.* II 2, 194a16 (δύο αἱ φύσεις), 22–7; 8, 199a30–2.

2 *Ph.* II 9, 200a7–15; *PA* I 1, 642a2–14; *Metaph.* VIII 4, 1044a27–9.

3 Cooper, "Hypothetical Necessity," 151–4, 158, argues that material causes act as the means required to realize a particular goal, namely the actualization of a certain formal cause. Thus, far from explaining the nature of whatever acts as a material cause, this kind of hypothetical necessity presupposes that the raw materials already have a nature of their own. In particular, the material elements and their powers are presupposed.

4 *Metaph.* VII 17, 1041a32–b11.

5 *Ph.* II 2, 194a36–b7; 7, 198a22–7; 9, 199b34–200a7.

6 *Ph.* II 2, 194a16, 22–7; 8, 199a30–2. Moreover, Aristotle frequently claims that natural science must look at both the material and the formal cause of natural substances: *Ph.* II 1, 193a28–b8; 2, 194a12–27; 7, 198a22–4; 8, 199a30–1; *Cael.* I 1, 268a1–7; *de An.* I 1, 403a24–b12; III 4, 429b13–14; *PA* I 1, 641a25–7; *GA* IV 4, 770b13–18; *Metaph.* II 3, 995a14–18; V 4, 1015a6–11; VI 1, 1025b34–1026a6; VIII 3, 1043a36–b4.

7 *Metaph.* VIII 4, 1044b1–3; IX 7, 1048b37–1049a18; XII 3, 1070a19–20.

8 *Metaph.* V 4, 1015a7–10; VIII 4, 1044a15–25; IX 7, 1049a18–27; *Ph.* II 1, 193a9–21.

9 *Ph.* VIII 4, 254b17–20.

10 *Cael.* I 12, 283b21–2; II 6, 288b15–19.

11 *de An.* II 7, 418b4–13.

12 *Ph.* II 1, 192b8–19.

13 *Ph.* II 1, 192b8–34, 193a29–30; 7, 198a27–b1; *Cael.* I 2, 268b16; *de An.* II 1, 412b15–17; *GA* II 1, 735a2–5; *Metaph.* V 4, 1015a13–15; VI 1, 1025b18–21; IX 8, 1049b5–10; XII 3, 1070a7–8.

14 *Ph.* II 8, 199a26–30. Such things seem to be natural by virtue of being products of a natural, internal principle of change, one located in the animals that make them. Presumably this derivative or paronymous sense of "nature" also applies to political communities, which Aristotle describes as existing by nature (*Pol.* I 2, 1253a2), although they too are not self-moving and are made by humans. Nests, webs, and political communities are also natural inasmuch as they serve the natural ends of the beings that make them.

15 *Ph.* II 1, 193a9–28.

16 *Ph.* II 1, 192b32–193b21.

17 At *Cael.* III 1, 298a27–b1, Aristotle discusses what is said to be by nature and again distinguishes between natural substances, such as the elements and the naturally occurring bodies made from them, and the natural functions, properties, and changes of such bodies.

18 *Metaph.* V 4, 1015a11–13.

19 *Ph.* II 1, 192b20–7.

20 *Ph.* II 1, 193a9–28.

21 *Ph.* II 1, 193a30–6.

22 *PA* I 1, 639b21–640a1; 640a10–27, b5–29.

23 *Ph.* II 1, 193a36–b12; 2, 194a12–27; 7, 198a21–33.

24 *Ph.* II 8, 199a30–2.

25 *Ph.* II 7, 198a22–4.

26 Aquinas, *De principiis naturae,* chap. 1, and *Expositio in libros Peri hermeneias* L. I, lect. IV; LeBlond, "Aristotle on Definition"; Owens, "Matter and

Predication," esp. 112; Charlton, *Aristotle's Physics I & II*, 75–7; Ackrill, "Aristotle's Definitions of *psuche*"; Gerson, "Artifacts, Substances, and Essences"; Kosman, "Animals and Other Beings in Aristotle"; Cohen, "Aristotle on Heat, Cold, and Teleological Explanation"; Gill, *Aristotle on Substance*, 146–67; Katayama, *Aristotle on Artifacts*; Frey, "Organic Unity and the Matter of Man"; Scharle, "Material and Efficient Causes in Aristotle's Natural Teleology."

27 *Ph.* II 1, 193a28–b5; 2, 194a21–7; 8, 199a12–15; *GC* II 9, 335b24–35; *de An.* II 1, 412b6–17; *PA* II 9, 654b27–33; *GA* I 18, 724a20–b1; *Metaph.* V 4, 1014b26–32; VII 7, 1032a12–22; 8, 1033b24–6; VIII 2, 1043a4–28.

28 *Ph.* I 7, 191a7–12. Charlton, *Aristotle's Physics*, 78–9, is correct that the point of this passage is to explain what an underlying principle is by generalizing from a number of paradigmatic examples. This interpretation is compatible with K. Cook's claim that it is more difficult to discern the underlying principles of natural substances than those of artefacts: "The Underlying Thing, the Underlying Nature and Matter: Aristotle's Analogy in *Physics* I 7," in *Nature, Knowledge and Virtue, Apeiron* 22, no. 4 (1989): 105–19. Indeed, it is for just this reason that Aristotle draws his examples of underlying principles from artefacts. See also *Ph.* II 1, 193a9–b12; 2, 194a12–27.

29 *Metaph.* IX 6, 1048a30–b4; he makes the same claim at *Metaph.* VIII 2, 1043a4–28.

30 J. Owens makes this point in order to argue that artefacts are bad examples to use in understanding the relation between the formal and material causes of natural substances: "The Aristotelian Argument for the Material Principle of Bodies," in *Naturphilosophie bei Aristoteles und Theophrast*, ed. I. Düring, 193–209 (Heidelberg: Stiehm Verlag, 1969). In response, see Cook, "The Underlying Thing, the Underlying Nature and Matter."

31 *Ph.* II 1, 193a28–b5; 2, 194a21–7; 8, 199a8–20; *GC* II 9, 335b24–35; *de An.* II 1, 412b6–17; *PA* I 1, 639b15–21; II 9, 654b27–33; *GA* I 18, 724a20–b1; 22, 730b5–32; II 1, 734b20–735a5; 4, 740b25–37; 5, 741b7–9; 6, 743a18–27, b20–5; V 8, 789b6–15; *Metaph.* V 4, 1014b26–32; VII 7, 1032a12–22; 8, 1033b24–6; IX 6, 1048a30–b4.

32 Most notably at *Ph.* II 8, 198b10–199b33; *de An.* II 4, 415b15–20; *PA* I 1, 639b12–640a20.

33 *Ph.* II 2, 194a33–b8; 8, 199a12–20; 9, 199b35–200a7; *PA* I 1, 639b15–19; *GA* II 1, 734b27–31.

34 *Ph.* II 1, 193a28–b12; 2, 194a12–27.

35 *Ph.* II 8, 199a20–1.

36 Aristotle repeatedly distinguishes between the formal causes of natural substances, on the one hand, and natural substances as composites of

a formal and material cause, on the other: *Metaph.* VII 7, 1032a15–22; 8, 1034a2–8; 17, 1041b11–19, 28–33; VIII 2, 1043a5–7, 12–28; 3, 1043b2–4, 18–21, and 1044a6–11; 6, 1045a7–10. If the material cause of a natural substance had no nature of its own, there would be no difference between the formal cause by itself and the composite substance containing both the formal and the material cause.

37 *Ph.* II 1, 192b13–15.

38 *de An.* II 4, 415b21–3; *Ph.* VIII 4, 255a5–10.

39 *GA* II 1, 735a13–18; *Metaph.* VII 7, 1032a24–5; XII 4, 1070b22–7. Even spontaneous generation, which is possible only for very simple entities, requires an external cause acting on the raw materials: *GA* III 11, 762b4–17; *Metaph.* VII 7, 1032b21–31; 9, 1034a9–b7.

40 *GA* I 18, 724a20–b1; 22, 730a33–b32.

41 *GC* I 4, 320a2–3; 7, 324b18; II 9, 335b24–35; *GA* I 18, 724b6–7; 22, 730b5–22; *Metaph.* IX 1, 1046a23.

42 *Ph.* II 1, 192b16–20.

43 S. Kelsey, "Aristotle's Definition of Nature," *Oxford Studies in Ancient Philosophy* 25 (2003): 59–87, argues that having an internal cause of change cannot be what Aristotle has in mind when he distinguishes natural substances from artefacts, because many natural changes are externally caused, beginning with generation. These latter, externally caused natural changes, however, are connected to the internally caused ones. In the first instance, the nature of a natural substance consists in its capacity for self-motion or self-change. The fundamental mode of causation in perceptible objects, however, is the transmission of properties through physical contact, which always involves reciprocal causation: every physical agent is affected at the same time that it moves something else. Thus, for every active capacity in a natural substance to cause motion or change, there must also be a corresponding passive capacity to be moved or changed. These passive capacities are also part of the nature of natural substances, grounded in their material cause, but this is a different sense of "nature" from the one Aristotle uses when distinguishing natural substances from artefacts. As we saw in the previous section, natural substances have many natures, beginning with their formal and material natures.

44 The expression Aristotle uses for the distinctive principle belonging to natural substances (ἀρχὴ κινήσεως) indicates that this principle is an efficient cause: *Ph.* II 1, 192b14, 21–2 and 193b4; 3, 194b29–30, 195a21–3; 7, 198a26; VIII 4, 254b14–17; *Cael.* I 2, 268b16; *PA* I 1, 639b13. At *Metaph.* IX 8, 1049b5–10, he puts the nature that natural substances have in the same genus as *dynamis*, inasmuch as both are efficient causes.

45 *Ph.* II 1, 192b8–32.
46 *GA* II 1, 735a3–5.
47 *Ph.* II 8, 199a20–1.
48 *Ph.* II 1, 193b8–12; 2, 194a12–17; 7, 198a24–7; *de An.* II 1, 412b10–17; 4, 415b8–28.
49 *Metaph.* V 6, 1016a4–5; VIII 2, 1043a2–28; 6, 1045a12–33; X 1, 1052a19–25.
50 *Ph.* II 9, 200a7–15; *de An.* II 4, 416a9–18; *PA* I 1, 639b21–640a1; 642a7–13; II 9, 655b12–13.
51 *Metaph.* VII 11, 1036a31–b8. At VIII 4, 1044a25–32, Aristotle distinguishes between artefacts that are different in kind, but have the same material cause, and artefacts where if the material cause is different, the formal cause must be different as well, because of its requirements for a certain kind of material cause. Even among artefacts, then, there is a difference in how wide or narrow the range of suitable raw materials is.
52 *Metaph.* VIII 5, 1045a3–6.
53 It also happens in the making of certain artefacts that their raw materials are greatly changed, e.g., the ingredients of a cake. The mark of an artefact, however, is not that it undergoes no natural changes. As we saw in section 7.3 above, there is something natural about every artefact, namely its raw materials and the natural motions that belong to them. The ingredients of the cake are hard to discern in the baked cake precisely because this part of the process is natural. The distinction between natural substances and artefacts is found at the point where an artificial formal cause is added to natural raw materials.
54 *Metaph.* V 6, 1015b34–1016a4; VIII 6, 1045a12–25; X 1, 1052a19–25.
55 Furth, *Substance, Form and Psyche*, 76–88, organizes natural substances and their parts into six different levels, with the lower levels acting as material causes for the higher levels: the material cause of the elements; the elements; uniform mixtures; the uniform parts of animals; the non-uniform parts of animals; and finally animals. The formal causes at the higher levels make demands on all of the lower-level material causes, right down to the elements.
56 *Ph.* I 7, 191a7–12; II 2, 194b9; *Metaph.* VII 10, 1035a8–9; IX 6, 1048a35–b9.
57 *GC* I 4, 320a2–3; II 9, 335b24–35; *GA* I 18, 724b6–7; IV 4, 770b15–17; *Metaph.* IX 1, 1046a23.
58 Frey, "Organic Unity and the Matter of Man," 193 and 196, argues against the existence of independent material causes in biological organisms by pointing out that the attributes of the material cause are usually insufficient to constitute the organism made from it. The argument for independent material causes, however, is not that they are sufficient

to explain the behaviour of biological organisms, only that they are necessary. See also Scharle, "Material and Efficient Causes in Aristotle's Natural Teleology," 29, who infers invalidly that the material cause of a natural substance is dependent on that substance's formal cause because the formal cause in question is indispensable to that substance. The indispensability of the formal cause, however, does not entail that it explains everything. She also mistakenly argues that the formal cause of a natural substance explains everything non-accidental about that substance, because the material cause of a natural substance has a formal cause of its own. While every material cause except prime matter does have a formal cause of its own, the formal cause of a material cause belongs to that material cause, not to the composite substance made from it: the formal cause of wood belongs to wood and is independent of the formal cause of the house made from that wood. The formal cause of a natural substance does not explain the formal causes of its raw materials, it presupposes them.

59 *Ph.* II 1, 193a28–b5; 2, 194a12–27; 9, 200a5–12; *de An.* II 4, 416a9–18; *GC* II 9, 335b24–35; *GA* I 1, 640b22–29; *Metaph.* IX 8, 1049b17–29.

Chapter Eight

1 *de An.* II 1, 412a15–21; 2, 414a17–22; III 4, 429b13–14; *GA* II 4, 738b20; *Metaph.* VII 10, 1035b27–31; 11, 1037a5–7; VIII 3, 1043a36–b4.

2 Aquinas, *De principiis naturae,* chap. 1; and Aquinas, *Expositio in libros Peri hermeneias* L. I, lect. IV; LeBlond, "Aristotle on Definition"; Owens, "Matter and Predication in Aristotle," esp. 112; Charlton, *Aristotle's Physics I & II,* 75–7; D. Balme, *Aristotle's De partibus animalium I and De generatione animalium I;* Ackrill, "Aristotle's Definitions of *psuche*"; Gerson, "Artifacts, Substances, and Essences"; Kosman, "Animals and Other Beings in Aristotle"; Cohen, "Aristotle on Heat, Cold, and Teleological Explanation"; Gill, *Aristotle on Substance,* 65–75, 146–67; Frey, "Organic Unity and the Matter of Man."

3 *Mete.* IV 12, 389b32–390a2, 10–14; *de An.* II 1, 412b18–22, 413a1; *PA* I 1, 640b34–641a6; *GA* I 19, 726b22–4; II 1, 734b25–31; *Metaph.* VII 10, 1035a31–3, b235; 11, 1036b30–2. Some parts are used for just one function, others for several. The teeth of certain animals, for example, are used for both eating and self-defence (*PA* III 1, 661b1–7); this dual use of a part, Aristotle suggests, occurs often in biological organisms (*PA* II 16, 659a22–3, 660a1; IV 10, 688a22–4).

4 *GA* II 1, 734b18–27; 735a13–26.

5 Ackrill, "Aristotle's Definitions of *psuche*," argues that the distinction between material and formal causes can be applied to artefacts, but not animals, because the proximate material cause of the latter is defined as a body with the capacity for life. As a result, the relevant material cause cannot "be picked out and (re-) identified in both an unformed and an in-formed state" (68). "The material in this case is *not* capable of existing *except* as the material of an animal, as matter *so in-formed*. The body we are told to pick out as the material 'constituent' of the animal depends for its very identity on its being alive, in-formed by *psyche*" (70).

6 *GC* I 4, 320a2–3; II 5, 332a17–18; 7, 334a23–4; *Ph.* I 9, 192a31–2; IV 9, 217a26–8; *Cael.* IV 5, 312a30–3; *Metaph.* XII 2, 1069b7–9.

7 Aquinas, *De principiis naturae,* chap. 1; and Aquinas, *Expositio in libros Peri hermeneias* L. I, lect. IV; LeBlond, "Aristotle on Definition"; Owens, "Matter and Predication," esp. 112; Charlton, *Aristotle's Physics I & II*, 75–7; Balme, *Aristotle's De partibus animalium I and De generatione animalium*; Ackrill, "Aristotle's Definitions of *psuche*"; Gerson, "Artifacts, Substances, and Essences"; Kosman, "Animals and Other Beings in Aristotle"; Cohen, "Aristotle on Heat, Cold, and Teleological Explanation"; Gill, *Aristotle on Substance*, 65–75, 146–67; Frey, "Organic Unity and the Matter of Man."

8 *de An.* II 1, 412a6–b5.

9 To this definition, he adds two more: the soul is the first actuality of a natural body with the potential for life, and the first actuality of a natural body that possesses organs. For our purposes, the differences between these three definitions are not important.

10 *de An.* II 1, 412b25–6.

11 *de An.* II 1, 412a11–15.

12 *de An.* II 1, 412b15–17.

13 This is how Hicks reads Aristotle's definition of the soul: Aristotle, *De anima*, translation and notes by R.D. Hicks (Cambridge: Cambridge University Press 1907), 308. He also discusses the ancient commentators who read Aristotle this way (309–11). More recently, this view was defended by W. Wehrle, "The Definition of Soul in Aristotle's *De anima* II 1 Is Not Analogous to the Definition of Snub," *Ancient Philosophy* 14 (1994): 297–317, esp. 310.

14 *de An.* II 1, 412b18–22.

15 *de An.* II 1, 412b20–7.

16 *Mete.* IV 12, 389b26–8; *PA* II 1, 646a13–24; *GA* I 1, 715a9–12; 18, 722a26–35; *Metaph.* XII 3, 1070a19.

17 Thus, Aristotle also applies his homonymy principle to the functional parts of artefacts: *Mete.* IV 12, 390a1, 10–14; *de An.* II 1, 412b14–15; *PA* I 1, 641a1–3.

18 *Mete.* IV 12, 389b32–390a2, 10–14; *de An.* II 1, 412b18–22, 413a1; *PA* I 1, 640b34–641a6; *GA* I 19, 726b22–4; II 1, 734b25–31; *Metaph.* VII 10, 1035a31–3, b23–5; 11, 1036b30–2.

19 To preserve the distinction between non-persisting functional parts and persisting material causes, some commentators have argued that Aristotle uses two different kinds of material cause in his account of biological organisms, sometimes called "functional matter" and "compositional matter"; functional matter acts as the proximate material cause of which biological organisms are made while they are alive, and compositional matter acts as the substratum persisting through generation and destruction: Hartman, *Substance, Body, and Soul* (1977), 113–15; B. Williams, "Hylomorphism," *Oxford Studies in Ancient Philosophy* 4 (1986): 189–99; T. Irwin, *Aristotle's First Principles* (Oxford: Clarendon, 1988), 240–2, 285–6; Cohen, "Hylomorphism and Functionalism"; J. Whiting, "Living Bodies," in Nussbaum and Rorty, *Essays on Aristotle's De anima*, 75–91; A. Code and J. Moravcsik, "Explaining Various Forms of Living"; C. Shields, "The Homonymy of the Body in Aristotle," *Archiv für Geschichte der Philosophie* 75 (1993), 1–30; F. Lewis, "Aristotle on the Relation between a Thing and Its Matter," in *Unity, Identity, and Explanation in Aristotle's Metaphysics* (1994), 247–77; C.V. Mirus, "Homonymy and the Matter of a Living Body," *Ancient Philosophy* 21 (2001): 357–73.

20 *de An.* II 4, 415b8–28.

21 *de An.* II 1, 413a1; 2, 413a26, b12 and 25; 4, 415a19, 416b17–19.

22 *de An.* I 1, 403a24–b19; 4, 408b11–15; II 2, 414a17–22; III 4, 429b13–14; 7, 431b12–16; *Metaph.* VII 10, 1035b27–31; 11, 1037a5–7; VIII 3, 1043a36–b4. At *GA* II 4, 738b20, he says that an animal is an ensouled body.

Chapter Nine

1 *APo.* II 11, 95a6–8; *Ph.* II 3, 194b32–195a3, 23–6; 7, 198b8–9; 8, 198b17; *GA* II 1, 731b23–4; V 8, 789b5–6.

2 *Metaph.* XIII 3, 1078a31–b2: here Aristotle says that mathematics does not deal with the good because the good belongs to activities, and the latter are not found in mathematical objects, but it does deal with the beautiful, or, more precisely, order, symmetry, and limit.

3 *Ph.* II 3, 195a8–11; *GC* I 7, 324b14–18; II 9, 335b20–9; *Metaph.* I 3, 983a30–2; VII 8, 1033b26–1034a5; 9, 1034b16–19; VIII 3, 1043b16–18; XII 6, 1071b14–16.

4 *Ph.* II 7, 198a24–b4; III 2, 202a3–9; VII 1, 242b59–63; 2, 243a34–5; *GA* II 4, 740b22–741a4.

5 *de An.* III 10, 433b10–21; *MA* 6, 700b23–701a2; *Metaph.* XII 7, 1072a26–7, b1–4.

6 *de An*. III 10, 433a17–26; *MA* 6, 700b35–701a2.

7 A. Woodfield, *Teleology* (Cambridge: Cambridge University Press, 1976), 206, argues that this is both Aristotle's position and the correct account of teleological explanation in its own right.

8 *Ph*. II 5, 196b10–13; 8, 198b34–6, 199b24–6.

9 *Ph*. II 5, 196b18–19; 8, 199a20–30, b26–33.

10 *Ph*. II 2, 194a29–30; 8, 199a8–20. There is much debate over just what a teleological explanation claims about such changes. Cooper, "Hypothetical Necessity," 151–4, argues that teleology primarily captures an instrumental, means-end relation. A. Gotthelf, "Aristotle's Conception of Final Causality," in *Philosophical Issues in Aristotle's Biology*, ed. A. Gotthelf and J. Lennox, 204–42 (Cambridge: Cambridge University Press, 1987), argues that teleology primarily applies to changes caused by a potentiality in the changing object that is irreducible to the potentialities for change found in the object's material causes. T. Scaltsas, "Commentary on Gotthelf," *Proceedings of the Boston Area Colloquium in Ancient Philosophy* 4 (1988): 140–7, argues that Aristotle is thinking primarily of the relation between a functional part and the whole of which it is a part. D. Charles, "Teleological Causation in the *Physics*," in *Aristotle's* Physics, ed. L. Judson, 101–28 (Oxford: Clarendon, 1991), argues that Aristotle's use of teleological explanation is inherently unclear, because he uses it to explain two very different kinds of change: the purposive behaviour of conscious agents and the proper functioning of internally complex objects. As we shall see below, there is also much debate about whether teleological explanation can be applied beyond individual substances to the interaction between different individuals, especially individuals belonging to different species.

11 *Ph*. II 2, 194a30–3.

12 More precisely, Aristotle distinguishes between the good that is the final cause of a change and the thing that benefits from that change: *Ph*. II 2, 194a34–6; *de An*. II 4, 415b2–3, 20–1; *Metaph*. XII 7, 1072b2–3; *EE* VII 15, 1249b15. Health, for example, is the final cause of healing, and the sick patient benefits from the healing.

13 *Ph*. II 7, 198a22–4 and b4–9; 8, 198b10–199b33, esp. 198b32–199a8; *de An*. II 4, 415b15–17; *PA* I 1, 639b12–640a20.

14 In addition to the texts cited in the previous note, see also *Cael*. I 4, 271a34; II 8, 290a31; *de An*. III 9, 432b21; 12, 434a31; *GA* I 1, 715b15–16.

15 Thus, not all of the regular motions and changes of natural substances must be explained teleologically, contrary to some commentators: Balme, "Greek Science and Mechanism"; and Balme, *Aristotle's De partibus animalium I and De generatione animalium I*; Ross, *Aristotle's Physics*,

43; Demos, "Structure of Substance According to Aristotle"; Owens, "Teleology of Nature in Aristotle"; Wieland, *Die aristotelische Physik*, 265. See Sorabji, *Necessity, Cause and Blame*, 144n3, for other commentators, both ancient and modern, who argue that for Aristotle every regular and non-accidental change in natural substances has a final cause of its own.

16 As we saw in section 1.2, Aristotle's definition of change is logically prior to his definition of natural change: *Ph.* III I, 200b15–25, b32–201a16, 34–b3.

17 *Ph.* II 2, 194a30–4.

18 *Metaph.* VIII 4, 1044b12.

19 Aristotle's distinction at *Ph.* II 5, 196b17–18, between things that happen for the sake of an end and things that do not implies that the latter category is not empty. See also *GA* V 1, 778a29–b2; 8, 789b19–22.

20 Some commentators see non-teleological physical necessity at work in the elements but deny that Aristotle extends this necessity to the natural substances made out of the elements, most notably, animate substances. Kosman, "Animals and Other Beings in Aristotle," argues that the elements, occurring on their own, exhibit what he calls Democritean necessity; in the case of animals, however, there is no such independent necessity arising from their material cause: "The body of an animal does not exhibit two beings, one qua matter and one qua what it is actually, for the actual being of that which is the matter just is its being qua matter" (378) "In the case of human being, … there is nothing that is human which is what it is independently of being human" (367). See also Gill, *Aristotle on Substance*, 145–67; Cohen, "Aristotle on Heat, Cold, and Teleological Explanation."

21 *Ph.* II 9, 200a1–5, 30–2; *PA* I 1, 642a2–b2; III 2, 663b13–24; *GA* I 4, 717a15–17; V 1, 778a29–b2, 12–19; 8, 789b19–22.

22 Other authors who argue for this position include Peck, "Preface" to Aristotle, *Generation of Animals* (Cambridge, MA: Loeb, 1942), xli–xlii; Mansion, *Introduction à la physique aristotélicienne*, 289–91; Charlton, *Aristotle's Physics I & II*, xvii and 120–3; Sorabji, *Necessity, Cause and Blame*, 149–54 and 162–3; Cooper, "Aristotle on Natural Teleology," esp. 210n8; and Cooper, "Hypothetical Necessity"; Friedman, "Matter and Necessity in *Physics* B 9, 200a15–30"; and Friedman, "Simple Necessity in Aristotle's Biology," *International Studies in Philosophy* 19 (1987): 1–9; Bradie and Miller Jr, "Teleology and Natural Necessity in Aristotle"; Leunissen, *Explanation and Teleology in Aristotle's Science of Nature*, esp. 47–8. See also C. Byrne, "Aristotle on Physical Necessity and the Limits of Teleological Explanation," *Apeiron* 35 (2002): 19–46.

23 *Ph.* II 5, 196b21–9, 197a8–21. The absence of final causes is particularly evident in spontaneous generation; on this topic, see Lennox, "Teleology,

Chance, and Aristotle's Theory of Spontaneous Generation"; and A. Gotthelf, "Teleology and Spontaneous Generation in Aristotle: A Discussion," in *Nature, Knowledge and Virtue, Apeiron* 22, no. 4 (1989): 181–93.

24 *Ph.* II 7, 198a24–6, b1–4; 8, 199a30–2; 9, 200a7–15, 34–5; *Mete.* IV 12, 390a4–5; *PA* I 1, 639b15–17, 640a16–19, 641a25–7; *GA* I 1, 715a4–6; V 1, 778b12–16.

25 *Ph.* II 7, 198b8–9.

26 *Ph.* II 1, 193b12–18; 7, 198b3–4; 9, 200a33–4; *GC* II 9, 335b5–7; *de An.* II 4, 415b8–21; *PA* I 1, 639b15–17, III 2, 663b23–4; *GA* II 4, 740b25–34; V 1, 778b6–7; *Metaph.* V 4, 1015a10–11; IX 8, 1050a4–7.

27 *Ph.* II 1, 193b8–12; 7, 198a26–7; *Mete.* IV 12, 389b26–390a12; *de An.* I 4, 408b11–15; II 1, 412a19–b6; 4, 415b21–8; *PA* III 2, 663b23–5; *GA* II 1, 732a1–5, 734a30–3; *Metaph.* IX 8, 1050a9–b6.

28 A similar position is defended by D. Henry, "Organismal Natures," in *Aristotle on Life*, edited by John Mouracade, *Apeiron* 41, no. 4 (2008): 47–74.

29 Balme, *Aristotle's De partibus animalium I and De generatione animalium I*, 79, is correct that if something results simply from physical necessity, there is nothing left for a final cause to explain. Thus, final causes can be used to explain only those features of perceptible objects that are over and above what can already be explained by physical necessity.

30 *PA* I 1, 639b14–16.

31 *Ph.* II 8, 198b16–199a8.

32 *Pol.* I 8, 1256b15–22. Artefacts also fit into this category inasmuch as their behaviour is good only insofar as it is beneficial for us.

33 For a recent defence of the internalist position, see Johnson, *Aristotle on Teleology*. See also Gotthelf, "Aristotle's Conception of Final Causality"; L. Judson, "Aristotelian Teleology," *Oxford Studies in Ancient Philosophy* 24 (2005): 341–66; R. Wardy, "Aristotelian Rainfall or the Lore of Averages," *Phronesis* 38 (1993): 18–30; Scharle, "Elemental Teleology in Aristotle's *Physics* 2.8."

34 D. Furley, "The Rainfall Example in *Physics* II.8," in *Cosmic Problems: Essays on Greek and Roman Philosophy of Nature*, 115–20 (Cambridge: Cambridge University Press, 1989); and D. Sedley, "Is Aristotle's Teleology Anthropocentric?" *Phronesis* 36 (1991): 179–96, defend an externalist account of this sort.

35 *APo.* II 11, 94b27–37.

36 *Ph.* II 8, 198b16–199a8.

37 *PA* I 1, 642a32–b4; II 14, 658b3–10; III 2, 663b23–5; IV 5, 679a25–30; *GA* II 4, 738a34–b9; 4, 739b22–31; 6, 743a37–b18; III 4, 755a22–6; IV 8, 776b29–35; V 8, 789a9–b8.

38 *Ph.* IV 8, 215a11–14, 19–22.
39 *APo.* I 9, 76a26–31.
40 *PA* II 9, 654b27–35; 655a11–14; 655b5–13; see also *de An.* II 4, 416a9–18; *PA* I 1, 642a9–14; III 2, 663b23–5; *GA* II 4, 738b1–9; 6, 743a21–7, 37–b18.
41 *PA* II 13, 657a31–3; *GA* V 1, 779b35–780a25.
42 *GA* V 1, 779b13–781a12.
43 *Ph.* II 9, 200a7–15; *PA* I 1, 639b21–30, 642a2–13, 32–b4; II 1, 646b15–19; *Metaph.* VIII 4, 1044a25–9.
44 Cooper, "Hypothetical Necessity," 158, makes a similar point when he argues that Aristotle's account of hypothetical necessity presupposes the material elements and their powers. He weakens the force of this claim, however, when he argues that, for Aristotle, simple material necessity "operates only against the background of hypothetical necessity ... [in] the formation and behaviour of living things," but not, for example, in the formation of ice on a pond (163). It is true that for Aristotle the simple necessity found in the material elements is generally insufficient to explain the formation and behaviour of biological organisms. Still, Aristotle explains many features of biological organisms by appealing to properties of the matter from which they are made, and this matter operates as it does of necessity, independent of whatever is made out of it. The converse of Cooper's claim, then, is equally true: the formation of biological organisms takes place against the background of simple physical necessity and only within the limits set by the latter.

Chapter Ten

1 *Ph.* I 7, 191a8–14; IV 2, 209b1–11; *Metaph.* IV 4, 1007b28–9; VII 3, 1029a20–6; VII 11, 1037a21–34; IX 7, 1049b2.
2 *Ph.* I 3, 186a28–34; 7, 190a14–17, b23–5, b36–191a3; 9, 192a3–6, 20–5; *GC* I 3, 319a29–b4; *Metaph.* VIII 1, 1042a25–b3; IX 7, 1049a27–36.
3 *Ph.* I 3, 186a28–34; 7, 190a14–17, b23–5, b36–191a3; 9, 192a3–6, 22–5; *GC* I 3, 319a29–b4.
4 *Metaph.* VII 16, 1040b5–10.
5 At *Metaph.* IX 7, 1049a27–36 and *Ph.* I 7, 191a7–14, Aristotle says that the material elements are not "thises" (τόδε τι), yet it is possible to state their nature; see also *Metaph.* VII 16, 1040b5–10; 17, 1041a20–b33. Sokolowski, "Matter, Elements and Substance in Aristotle," claims that only those things possessing a formal cause that makes them to be one and unified, as opposed to a heap, can be said to be substances in actuality. On this view, the material elements are only potentially substances because, as such,

they lack this kind of unity. Nevertheless, Aristotle also claims that to be a substance is to possess a kind of independent existence that properties lack, and the material elements clearly possess this latter kind of actuality. This ambiguity in the notion of actuality points to the fact that Aristotle is using two quite distinct criteria to decide what counts as a substance: functional indivisibility and unity, on the one hand, and independence in existence, on the other. When applied individually, they point to two very different sorts of thing as substances. At the very least, it must be recognized that the actuality that the material elements lack presupposes another kind of actuality that the elements must have in their own right if they are to receive the actuality of unified substances.

6 This account of potentiality and actuality is borne out by what Aristotle says at *Metaph.* IX 1, 1045b32–1046a4; 6, 1048a25–b9; 8, 1050a15. When it comes to setting out the second, metaphysically more important sense of "actuality" (ἐνέργεια) and "completion" (ἐντελέχεια), he says they cannot be defined directly but must be generalized from certain examples. All of the examples he then gives involve the relation between raw materials and the physical artefact made from those materials; as we have seen, the raw materials of every finished artefact have a determinate nature of their own that persists in the artefact made from them.

7 At *Physics* II 3, 195b4–21, Aristotle says that potentiality and actuality are said of all four causes, i.e., one can speak of both a potential and an actual material cause and a potential and an actual formal cause. Thus, one cannot simply equate potentiality with the material cause and actuality with the formal cause. As Dye, "Aristotle's Matter as a Sensible Principle," 79, and Graham, "Paradox of Prime Matter," 483, argue, Aristotle's claim that actuality is prior to potentiality means that every potentiality presupposes some actuality.

8 *GC* I 4, 320a2–3; 7, 324b18; II 9, 335b24–35; *GA* I 18, 724b6–7; IV 4, 770b15–17; *Metaph.* IX 1, 1046a23.

9 *Metaph.* V 2, 1013a26–9; VII 7, 1032b1–2, 14; 8, 1033b5–8; 10, 1035a1–14; 1035b4–16, 32; 1036a22–3; 11, 1036a26–1037a20; VIII 3, 1043b1–2.

10 *Metaph.* VII 17, 1041a6–b4.

11 This claim is not refuted by pointing out that Aristotle sometimes speaks of a second definition of perceptible objects that does include their material cause, analogous to the definition of a snub nose (*Ph.* II 2, 194a12–17; *de An.* I 1, 403a24–b19; III 4, 429b13–14; *Cael.* I 9, 277b32–278b9; *Metaph.* VI 1, 1025b28–1026a6; VII 10, 1034b34–1035a7; 11, 1036a26–b2, 1037a1–5). For if the essence of a perceptible object is captured by this kind of dual definition, then perceptible objects still have two natures, corresponding to the two parts of their definition.

12 *Metaph.* VI 2, 1026b31–3.
13 Gerson, "Artifacts, Substances, and Essences," 55, says that a composite substance is "reductively identical with its form," a feature that artefacts do not have. Gill, *Aristotle on Substance*, 167, says that the nature of a generated substance is "exhausted by its form."
14 *Metaph.* VII 16, 1040b25–6.
15 *Ph.* IV 1, 208b22–5; 4, 212a6–7; 5, 212b28–9.
16 *Ph.* I 7, 190b10–13.
17 *Ph.* I 7, 189b30–190a13; V 1, 224b1–7; *GC* I 3, 319b2–4.
18 *Ph.* I 7, 190a14–17, 190b10–17, 22–5, 33–191a3; 9, 192a3–6; *Metaph.* VII 8, 1033b12–13; *GC* I 3, 319a29–b2; 5, 320b12–14.
19 *Metaph.* IV 1, 1003a21–32; 2, 1003b15–19; 3, 1005a19–b2; *APo.* I 2, 72a15–19; 9, 76a17–19; 11, 77a27–9.

Works Cited

Ackrill, J.L. "Aristotle's Definitions of *psuche.*" In Barnes, Schofield, and Sorabji, *Articles on Aristotle*, 4:65–75.

Allen, R.E. "Substance & Predication in Aristotle's *Categories.*" In *Exegesis and Argument*, edited by E.N. Lee, P.D. Mourelatos, and R.M. Rorty. Supplement, *Phronesis*, no. 1 (Assen: Van Gorcum, 1973), 362–73.

Anders, J. "*Problēmata Mēchanika*, the Analytics, and Projectile Motion." *Apeiron* 46 (2013): 106–35.

Aquinas, T. *De principiis naturae* (Rome: Leonine Edition, 1976), 43:37–47.

– *Expositio in libros Peri hermeneias*. Edited by R.M. Spiazzi (Turin: Marietti, 1955), 1–144.

– *Questiones disputatae de anima*. Vol. 24/1 (Rome: Leonine Edition: 1996).

– *Questiones disputatae de potentia*. Edited by R.M. Spiazzi (Turin: Marietti, 1952), 313–85.

– *Summa theologiae*. Institute of Mediaeval Studies Edition. (Ottawa: Commissio Piana, 1953), 5 vols.

– *Super libros De generatione et corruptione*. In *Sancti Thomae Aquinatis doctoris angelici Opera omnia iussu Leonis XII*. Edited by the Order of Preachers (Rome: Leonine Edition, 1882–), 3:261–322.

Balme, D.M. *Aristotle's De partibus animalium I and De generatione animalium I* (Oxford: Clarendon, 1972).

– "Greek Science and Mechanism: 1. Aristotle on Nature and Chance." *Classical Quarterly* 33 (1939): 129–38.

– "Teleology and Necessity." In *Philosophical Issues in Aristotle's Biology*, edited by A. Gotthelf and J. Lennox, 275–85 (Cambridge: Cambridge University Press, 1987).

Barnes, J., M. Schofield, and R. Sorabji, eds. *Articles on Aristotle.* Vols 1–4 (London: Duckworth, 1975–9).

Berryman, S. *The Mechanical Hypothesis in Ancient Greek Natural Philosophy* (Cambridge: Cambridge University Press, 2009).

Bodnar, I. "Movers and Elemental Motions in Aristotle." *Oxford Studies in Ancient Philosophy* 15 (1997): 81–117.

Boehm, R. *Das Grundlegende und das Wesentliche* (Den Haag: Nijhoff, 1965).

Bogaard, P. "Heaps or Wholes: Aristotle's Explanation of Compound Bodies." *Isis* 70 (1979): 11–29.

Bogen, J. "Fire in the Belly: Aristotelian Elements, Organisms, and Chemical Compounds." *Pacific Philosophical Quarterly* 76 (1995): 370–404.

Bradie, M., and F. Miller Jr. "Teleology and Natural Necessity in Aristotle." *History of Philosophy Quarterly* 1 (1984): 133–45.

Burnyeat, M.F. "Is an Aristotelian Philosophy of Mind Still Credible?" In Nussbaum and Rorty, *Essays on Aristotle's De anima*, 15–26.

Butterfield, H. "The Historical Importance of a Theory of Impetus." In his *Origins of Modern Science*, rev. ed., 13–28 (London: G. Bell, 1957).

Byrne, C. "Aristotle on Physical Necessity and the Limits of Teleological Explanation." *Apeiron* 35 (2002): 19–46.

– "Compositional & Functional Matter: Aristotle on the Material Cause of Biological Organisms." *Apeiron* 48 (2015): 387–406.

– "Matter and Aristotle's Material Cause." *Canadian Journal of Philosophy* 31 (2001): 85–112.

– "Prime Matter and Actuality." *Journal of the History of Philosophy* 33 (1995): 197–224.

Carteron, H. "Does Aristotle Have a Mechanics?" In Barnes, Schofield, and Sorabji, *Articles on Aristotle*, 1:161–74.

Charles, D. "Teleological Causation in the *Physics*." In *Aristotle's* Physics, edited by L. Judson, 101–28 (Oxford: Clarendon, 1991).

Charlton, W. *Aristotle's Physics I & II* (Oxford: Clarendon, 1970).

– "Prime Matter: A Rejoinder." *Phronesis* 28 (1983): 197–211.

Code, A. "The Persistence of Aristotelian Matter." *Philosophical Studies* 29 (1976): 357–67.

– "Potentiality in Aristotle's Science and Metaphysics." *Pacific Philosophical Quarterly* 76 (1995): 405–18.

Code, A., and J. Moravcsik. "Explaining Various Forms of Living." In Nussbaum and Rorty, *Essays on Aristotle's De anima*, 129–45.

Cohen, I.B. *The Birth of a New Physics*. 2nd ed. (New York: Norton, 1985).

– "Quantum in se est: Newton's Concept of Inertia in Relation to Descartes and Lucretius." *Notes and Records of the Royal Society of London* 19 (1964): 131–55.

Cohen, S. "Aristotle's Doctrine of the Material Substrate." *Philosophical Review* 93 (1984): 171–94.

– "Aristotle on Heat, Cold, and Teleological Explanation." *Ancient Philosophy* 9 (1989): 255–70.

Cohen, S.M. "Hylomorphism and Functionalism." In Nussbaum and Rorty, *Essays on Aristotle's De anima*, 57–73.

Cook, K. "The Underlying Thing, the Underlying Nature and Matter: Aristotle's Analogy in *Physics* I 7." In *Nature, Knowledge and Virtue*, edited by T. Penner and R. Kraut, *Apeiron* 22, no. 4 (1989): 105–19.

Cooper, J. "Aristotle on Natural Teleology." In *Language and Logos*, edited by M. Schofield and M. Nussbaum, 197–222 (Cambridge: Cambridge University Press, 1982).

– "Hypothetical Necessity." In *Aristotle on Nature and Living Things*, edited by A. Gotthelf, 151–67 (Pittsburgh: Mathesis Publications, 1985).

Demos, R. "The Structure of Substance According to Aristotle." *Philosophy & Phenomenological Research* 5 (1944–5): 255–68.

Distelzweig, P. "The Intersection of the Mathematical and Natural Sciences: The Subordinate Sciences in Aristotle." *Apeiron* 46 (2013): 85–105.

Drabkin, I.E. "Notes on the Laws of Motion in Aristotle." *American Journal of Philology* 59 (1938): 60–84.

Düring, I. *Aristotle's Chemical Treatise: Meteorologica* IV (Göteborg: Elanders, 1944).

Dye, J.W. "Aristotle's Matter as a Sensible Principle." *International Studies in Philosophy* 10 (1978): 59–84.

Falcon, A. *Aristotle and the Science of Nature* (Cambridge: Cambridge University Press, 2005).

Fine, K. "Aristotle on Matter." *Mind* 101 (1992): 37–57.

– "The Problem of Mixture." *Pacific Philosophical Quarterly* 76 (1995): 266–369.

FitzGerald, J.J. "'Matter' in Nature and the Knowledge of Nature: Aristotle and the Aristotelian Tradition." In McMullin, *The Concept of Matter*, 79–98.

Frede, M. "On Aristotle's Conception of the Soul." In Nussbaum and Rorty, *Essays on Aristotle's De anima*, 93–107.

Freeland, C. "Aristotle on Bodies, Matter, and Potentiality." In *Philosophical Issues in Aristotle's Biology*, edited by A. Gotthelf and J. Lennox, 392–407 (Cambridge: Cambridge University Press, 1987).

Freudenthal, G. *Aristotle's Theory of Material Substance: Heat and Pneuma, Form and Soul* (Oxford: Clarendon, 1995).

Frey, C. "Organic Unity and the Matter of Man." *Oxford Studies in Ancient Philosophy* 32 (2007): 167–204.

Friedman, R. "Matter and Necessity in *Physics* B 9, 200a15–30." *Ancient Philosophy* 3 (1983): 8–11.

– "Simple Necessity in Aristotle's Biology." *International Studies in Philosophy* 19 (1987): 1–9.

Furley, D. "The Rainfall Example in *Physics* II.8." In his *Cosmic Problems: Essays on Greek and Roman Philosophy of Nature*, 115–20 (Cambridge: Cambridge University Press, 1989).

Furth, M. *Substance, Form and Psyche: An Aristotelian Metaphysics* (Cambridge: Cambridge University Press, 1988).

Gerson, L. "Artifacts, Substances, and Essences." *Apeiron* 18 (1984): 50–8.

Gill, M.L. *Aristotle on Substance: The Paradox of Unity* (Princeton: Princeton University Press, 1989).

Gotthelf, A. "Aristotle's Conception of Final Causality." In *Philosophical Issues in Aristotle's Biology*, edited by A. Gotthelf and J. Lennox, 204–42 (Cambridge: Cambridge University Press, 1987).

– "Darwin on Aristotle." *Journal of the History of Biology* 32 (1999): 3–30.

– "Teleology and Spontaneous Generation in Aristotle: A Discussion." In *Nature, Knowledge and Virtue*, edited by T. Penner and R. Kraut, *Apeiron* 22, no. 4 (1989): 181–93.

Graham, D. "The Paradox of Prime Matter." *Journal of the History of Philosophy* 25 (1987): 475–90.

Granger, H. "Deformed Kinds and the Fixity of Species." *Classical Quarterly* 37 (1987): 110–16.

Grene, M. *A Portrait of Aristotle* (London: Faber & Faber, 1963).

Hankinson, R.J. *Cause and Explanation in Ancient Greek Thought* (New York: Oxford University Press, 1998).

– "Chapter 5: Science." In *The Cambridge Companion to Aristotle*, edited by J. Barnes, 140–67 (Cambridge: Cambridge University Press, 1995).

Hartman, E. *Substance, Body, and Soul: Aristotelian Investigations* (Princeton: Princeton University Press, 1977).

Heath, T. *Mathematics in Aristotle* (Oxford: Clarendon, 1949).

Heinaman, R. "Is Aristotle's Definition of Change Circular?" *Apeiron* 27 (1994): 25–37.

Henry, D. "Organismal Natures." In *Aristotle on Life*, edited by John Mouracade. *Apeiron* 41, no. 4 (2008): 47–74.

– "Substantial Generation in Physics I 5–7." In *Aristotle's* Physics*: A Critical Guide*, edited by M. Leunissen, 144–61 (Cambridge: Cambridge University Press, 2015).

Hicks, R.D. "Translation and Notes." In Aristotle, *De anima*, 173–588 (Cambridge: Cambridge University Press, 1907).

Hussey, E. "Aristotle's Mathematical Physics: A Reconstruction." In *Aristotle's* Physics*: A Collection of Essays*, edited by L. Judson, 213–42 (New York: Clarendon, 1995).

– *Aristotle's Physics Books III & IV* (Oxford: Clarendon, 1983).

Irwin, T. *Aristotle's First Principles* (Oxford: Clarendon, 1988).
Jaeger, W. *Aristoteles* (Berlin: Weidmann, 1923).
Jammer, M. *Concepts of Force: A Study in the Foundations of Dynamics* (Cambridge, MA: Harvard University Press, 1957).
– *Concepts of Mass in Classical and Modern Physics* (Cambridge, MA: Harvard University Press, 1961).
Joachim, H.H. "Commentary." In Aristotle, *De generatione et corruptione,* 62–277 (Oxford: Clarendon, 1922).
Johnson, M. *Aristotle on Teleology* (Oxford: Oxford University Press, 2005).
Jones, B. "Aristotle's Introduction of Matter." *Philosophical Review* 83 (1974): 474–500.
Judson, L. "Aristotelian Teleology." *Oxford Studies in Ancient Philosophy* 29 (2005): 341–66.
Katayama, E. *Aristotle on Artifacts: A Metaphysical Puzzle* (Albany, NY: SUNY Press, 1999).
Kelsey, S. "Aristotle's Definition of Nature." *Oxford Studies in Ancient Philosophy* 25 (2003): 59–87.
King, H.R. "Aristotle without *Materia Prima*." *Journal of the History of Ideas* 17 (1956): 370–89.
Kosman, L.A. "Animals and Other Beings in Aristotle." In *Philosophical Issues in Aristotle's Biology*, edited by A. Gotthelf and J. Lennox, 360–91 (Cambridge: Cambridge University Press, 1987).
Kuhn, T. "Mathematical versus Experimental Traditions in the Development of Physical Science." In *The Essential Tension*, 31–65 (Chicago: University of Chicago Press, 1977).
– "Preface." In *The Essential Tension*, xi–xii (Chicago: University of Chicago Press, 1977).
– "What Are Scientific Revolutions?" In *The Probabilistic Revolution*. Vol. 1, *Ideas in History*, edited by L. Kruger, L.J. Daston, and M. Heidelberger, 7–22 (Cambridge, MA: MIT Press, 1987); reprinted in Kuhn's *The Road since Structure*, edited by J. Conant and J. Haugeland, 13–32 (Chicago: University of Chicago Press, 2000).
Kung, J. "Can Substance Be Predicated of Matter?" *Archiv für Geschichte der Philosophie* 60 (1978): 140–59.
LeBlond, J.-M. "Aristotle on Definition." In Barnes, Schofield, and Sorabji, *Articles on Aristotle*, 3:63–79.
Lee, H.D.P. "Preface to Second Edition" and "Introduction." In Aristotle, *Meteorologica*, xiii–xxi (Cambridge, MA: Loeb, 1962).
Lennox, J. "Material and Formal Natures in Aristotle's *De partibus animalium*." In *Aristotelische Biologie: Intentionen, Methoden, Ergebnisse*, edited by

W. Kullmann and S. Föllinger, 163–81 (Stuttgart: Steiner, 1997); reprinted in J. Lennox, *Aristotle's Philosophy of Biology: Studies in the Origins of Life Science*, 182–204 (Cambridge: Cambridge University Press, 2001).

– "Teleology, Chance, and Aristotle's Theory of Spontaneous Generation." *Journal of the History of Philosophy* 20 (1982), 219–38; reprinted in J. Lennox, *Aristotle's Philosophy of Biology: Studies in the Origins of Life Science*, 229–49 (Cambridge: University of Cambridge Press, 2001).

Leroi, A.M. *The Lagoon: How Aristotle Invented Science* (New York: Viking, 2014).

Leunissen, M. "Aristotle on Natural Character and Its Implications for Moral Development." *Journal of the History of Philosophy* 50 (2012): 507–30.

– *Explanation and Teleology in Aristotle's Science of Nature* (Cambridge: Cambridge University Press, 2010).

Leunissen, M., and A. Gotthelf. "What's Teleology Got to Do with It? A Reinterpretation of Aristotle's *Generation of Animals* V." *Phronesis* 55 (2010): 325–56; reprinted in *Teleology, First Principles, and Scientific Method in Aristotle's Biology*, edited by A. Gotthelf, 108–30 (Oxford: Oxford University Press, 2012).

Lewis, F. "Aristotle on the Relation between a Thing and Its Matter." In *Unity, Identity, and Explanation in Aristotle's Metaphysics*, edited by T. Scaltsas, D. Charles, and M.L. Gill, 247–77 (Oxford: Oxford University Press, 1994).

Lobkowicz, N. "Discussion." In McMullin, *The Concept of Matter*, 120.

Luyten, N. "Matter as Potency." In McMullin, *The Concept of Matter*, 122–33.

Makin, S. "Commentary." In Aristotle, *Metaphysics Theta*, 17–269 (Oxford: Clarendon, 2006).

Mansion, A. *Introduction à la physique aristotélicienne*. 2nd ed. (Louvain: Institut supérieur de philosophie, 1946).

Mansion, S. "The Ontological Composition of Sensible Substances in Aristotle (*Metaphysics* Z 7–9)." In Barnes, Schofield, and Sorabji, *Articles on Aristotle*, 3:80–7.

Matthen, M. "The Four Causes in Aristotle's Embryology." In *Nature, Knowledge and Virtue*, edited by T. Penner and R. Kraut, *Apeiron* 22, no. 4 (1989): 159–79.

McMullin, E., ed. *The Concept of Matter* (Notre Dame: University of Notre Dame Press, 1963).

– "Matter as a Principle." In McMullin, *The Concept of Matter*, 169–208.

Mirus, C.V. "Homonymy and the Matter of a Living Body." *Ancient Philosophy* 21 (2001): 357–73.

Morison, B. *On Location: Aristotle's Concept of Place* (Oxford: Clarendon, 2002).

Nussbaum, M., and A. Rorty, eds. *Essays on Aristotle's De anima* (Oxford: Clarendon, 1992).

Owen, G.E.L. "Aristotelian Mechanics." In *Aristotle on Nature and Living Things*, edited by A. Gotthelf, 227–45 (Pittsburgh: Mathesis Publications, 1985).

Owens, J. "The Aristotelian Argument for the Material Principle of Bodies." In *Naturphilosophie bei Aristoteles und Theophrast*, edited by I. Düring, 193–209 (Heidelberg: Stiehm Verlag, 1969).

– "Matter and Predication in Aristotle." In McMullin, *The Concept of Matter*, 99–115; reprinted in *Aristotle: The Collected Papers of Joseph Owens*, edited by J. Catan, 35–47 (Albany, NY: SUNY Press, 1981).

– "The Teleology of Nature in Aristotle." *Monist* 52 (1968): 159–73; reprinted in *Aristotle: The Collected Papers of Joseph Owens*, edited by J. Catan, 136–47 (Albany, NY: SUNY Press, 1981).

Page, C. "Predicating Forms of Matter in Aristotle's *Metaphysics*." *Review of Metaphysics* 39 (1985): 57–82.

Peck, A.L. "Preface" to Aristotle, *Generation of Animals*, v–xxxvii (Cambridge, MA: Loeb, 1942).

Rist, J. *The Mind of Aristotle* (Toronto: University of Toronto Press, 1989).

Robinson, H.M. "Prime Matter in Aristotle." *Phronesis* 19 (1974): 168–88.

Ross, W.D. *Aristotle*. 5th ed. (London: Methuen, 1949).

– *Aristotle's Physics* (Oxford: Oxford University Press, 1936).

Sambursky, S. *The Physical World of the Greeks* (London: Routledge & Paul, 1956; ppb. edition: Princeton: Princeton University Press, 1987).

Sarton, G. *A History of Science*. Vol. 1 (Cambridge, MA: Harvard University Press, 1952).

Scaltsas, T. "Commentary on Gotthelf." *Proceedings of the Boston Area Colloquium in Ancient Philosophy* 4 (1988): 140–7.

– "Substratum, Subject, and Substance." *Ancient Philosophy* 5 (1985): 215–40.

Scharle, M. "Elemental Teleology in Aristotle's *Physics* 2.8." *Oxford Studies in Ancient Philosophy* 34 (2008): 147–83.

– "Material and Efficient Causes in Aristotle's Natural Teleology." In *Aristotle on Life*, edited by J. Mouracade, *Apeiron* 41, no. 4 (2008): 27–45.

Sedley, D. "Is Aristotle's Teleology Anthropocentric?" *Phronesis* 36 (1991): 179–96.

Shields, C. "The Homonymy of the Body in Aristotle." *Archiv für Geschichte der Philosophie* 75 (1993): 1–30.

Sokolowski, R. "Matter, Elements and Substance in Aristotle." *Journal of the History of Philosophy* 8 (1970): 263–88.

Solmsen, F. "Aristotle and Prime Matter: A Reply to Hugh R. King." *Journal of the History of Ideas* 19 (1958): 243–52.

– "Aristotle's Word for 'Matter.'" In *Didascaliae*, edited by S. Prete, 393–408 (New York: B. Rosenthal, 1961).

Sorabji, R. *Matter, Space, and Motion: Theories in Antiquity and Their Sequel* (Ithaca, NY: Cornell University Press, 1988).

– *Necessity, Cause and Blame* (London: Duckworth 1980).

Suppes, P. "Aristotle's Concept of Matter and Its Relation to Modern Concepts of Matter." *Synthese* 28 (1974): 27–50.

Wardy, R. "Aristotelian Rainfall or the Lore of Averages." *Phronesis* 38 (1993): 18–30.

Waterlow, S. *Nature, Change, and Agency in Aristotle's Physics* (Oxford: Clarendon, 1982).

Wedin, M. *Aristotle's Theory of Substance* (Oxford: Oxford University Press, 2002).

Wehrle, W. "The Definition of Soul in Aristotle's *De anima* II 1 Is Not Analogous to the Definition of Snub." *Ancient Philosophy* 14 (1994): 297–317.

Westfall, R. *The Construction of Modern Science: Mechanisms and Mechanics* (Cambridge: Cambridge University Press, 1977).

White, M.J. "Concepts of Space in Greek Thought." *Apeiron* 29 (1996): 183–98.

Whiting, J. "Living Bodies." In Nussbaum and Rorty, *Essays on Aristotle's De anima*, 75–91.

Wieland, W. *Die aristotelische Physik.* 2nd ed. (Göttingen: Vandenhoeck & Ruprecht, 1970).

Williams, B. "Hylomorphism." *Oxford Studies in Ancient Philosophy* 4 (1986): 189–99.

Williams, C.J.F. *Aristotle's De generatione et corruptione* (Oxford: Clarendon, 1982).

Witt, C. "Aristotle on Deformed Kinds." *Oxford Studies in Ancient Philosophy* 43 (2012): 83–106.

Wolter, A. "The Ockhamist Critique." In McMullin, *The Concept of Matter*, 144–66.

Woodfield, A. *Teleology* (Cambridge: Cambridge University Press, 1976).

Index of Texts from Aristotle

www.ingramcontent.com/pod-product-compliance
Lightning Source LLC
LaVergne TN
LVHW041113090826
844660LV00061B/870/J

* 9 7 8 1 4 8 7 5 0 3 9 6 3 *